MW00892545

A Level Physics

Multiple Choice Questions and Answers (MCQs)

Quiz & Practice Tests with Answer Key

By Arshad Iqbal

Table of Contents

Chapter 1

Accelerated Motion MCQs

MCQ 1: Only force acting on a bouncing ball is

A. gravity

B. weight of ball

C. friction

D. both A and B

MCQ 2: Accelerometer detects the

A. small acceleration

B. large acceleration

C. small deceleration

D. large acceleration and deceleration

MCQ 3: If the gradient of a graph is negative, then the acceleration is

A. positive

B. negative

C. zero

D. 1

MCQ 4: If a student drops a stone from a cliff of height 30 m and the time it takes to reach the ground is 2.6 s, then the acceleration due to gravity is

A. 9 ms^{-2}

B. 10 ms^{-2}

C. 4 ms^{-2}

D. 8.8 ms^{-2}

MCQ 5: Gradient of line of velocity-time graph is tells us the

A. velocity

B. acceleration

C. distance

D. time

MCQ 6: A stone is thrown upwards with initial velocity of 20 ms^{-1}, the height that stone will reach would be

A. 20 m

B. 30 m

C. 40 m

D. 50 m

MCQ 7: Projectile will attain its maximum range, if it is fired at an angle of

A. 30°

B. 47°

C. 90°

D. 45°

MCQ 8: Horizontal component of a bouncing ball is

A. affected by gravity

B. unaffected by gravity

C. affected by weight

D. affected by contact force

MCQ 9: When ball having a projectile motion is rising up, it

A. decelerates

B. accelerates

C. rises up with constant acceleration

D. acceleration becomes zero

MCQ 10: Equation of motion can be used for

A. straight line motion only

B. curved motion only

C. motion along the circular path

D. all types of motion

MCQ 11: Acceleration of free fall depends on the

A. surface

B. weight of object

C. distance from center of Earth

D. size of object

MCQ 12: If initial velocity of an object is zero, then distance covered by it in time t and acceleration of 9.8 ms^{-2} would be

A. 2.9 t^2

B. 3 t^2

C. 4 t^2

D. 4.9 t^2

MCQ 13: As the ball falls towards the ground, its velocity

A. increases

B. decreases

C. remains constant

D. becomes zero

MCQ 14: Gradient of velocity-time graph tells us about object's

A. velocity

B. displacement

C. distance

D. acceleration

MCQ 15: An object whose velocity is changing is said to be in a state of

A. acceleration

B. rest

C. equilibrium

D. Brownian motion

MCQ 16: Acceleration of train when it is moving steadily from 4.0 ms^{-1} to 20 ms^{-1} in 100 s is

A. 1 ms^{-2}

B. 2 ms^{-2}

C. 0.16 ms^{-2}

D. 3 ms^{-2}

MCQ 17: If we get a straight line with positive slope then its acceleration is

A. increasing

B. decreasing

C. zero

D. constant

MCQ 18: If a spinster staring from rest has acceleration of 5 ms^{-2} during 1st 2.0 s of race then her velocity after 2 s is

A. 20 ms^{-1}

B. 10 ms^{-1}

C. 15 ms^{-1}

D. 5 ms^{-1}

MCQ 19: Horizontal distance travelled by a ball if it's thrown with initial velocity of 20 ms^{-1} at an angle of 30° is

A. 24 m

B. 56 m

C. 35.3 m

D. 36.3 m

MCQ 20: If a car starting from rest reaches a velocity of 18 ms^{-1} after 6.0 s then its acceleration is

A. 1 ms^{-2}

B. 2 ms^{-2}

C. 3 ms^{-2}

D. 4 ms^{-2}

MCQ 21: A train travelling at 20 ms^{-1} accelerates at 0.5 ms^{-2} for 30 s, the distance travelled by train is

A. 825 m

B. 700 m

C. 650 m

D. 600 m

MCQ 22: Area under velocity-time graph tells us the

A. time

B. acceleration

C. displacement

D. velocity

Chapter 2

Alternating Current MCQs

MCQ 1: If a secondary coil has 40 turns, and, a primary coil with 20 turns is charged with 50 V of potential difference, then the potential difference in the secondary coil would be

A. 50 V in secondary coil

B. 25 V in secondary coil

C. 60 V in secondary coil

D. 100 V in secondary coil

MCQ 2: Generators at a power station produce electric power at voltage

A. 45 kW

B. 50 kW

C. 60 kW

D. 25 kW

MCQ 3: Equation which measures alternating voltage is

A. $V\sin \omega$

B. $\sin t$

C. $V_o \sin\omega t$

D. V=IR

MCQ 4: In transformer, alternating current is induced in

A. primary coil

B. secondary coil

C. iron core

D. resistor

MCQ 5: High voltages lead to

A. less power loss

B. more power loss

C. high current

D. high resistance

MCQ 6: Graph of alternating current is a

A. cos wave

B. tan wave

C. curve

D. sine wave

MCQ 7: A component that allows only unidirectional current to pass through it is

A. resistor

B. inductor

C. transformer

D. diode

MCQ 8: Average power dissipated in resistor if sinusoidal p.d of peak value 25 V is connected across a 20 Ω resistor is

A. 15.6 W

B. 15 W

C. 16 W

D. 17 W

MCQ 9: Amount of DC voltage as compare to V_o is

A. 50%

B. 60%

C. 70%

D. 80%

MCQ 10: A well designed transformer loses power under

A. 10%

B. 1.20%

C. 0.10%

D. 20%

MCQ 11: Ratio of voltages is equal to ratio of

A. iron sheets in core

B. coil

C. number of turns in coil

D. all of above

MCQ 12: Process in which AC is converted into DC is called

A. induction

B. rectification

C. inversion

D. dispersion

MCQ 13: Steady DC voltage is also known as

A. square velocity

B. root mean square velocity

C. root velocity

D. velocity

MCQ 14: Ratio of voltages is inverse ratio of the

A. transformer

B. resistor

C. iron core

D. current

MCQ 15: If output voltage is greater than input voltage, then transformer is

A. step up

B. step down

C. faulty

D. fatal

MCQ 16: Highest point on AC graph is known as

A. peak value

B. amplitude

C. frequency

D. wave front

Chapter 3

AS Level Physics MCQs

MCQ 1: There are three equations of uniformly accelerated motion, the odd one out is

A. final_velocity = initial_velocity + (acceleration × time)

B. distance_moved = (initial_velocity × time) + (0.5 × acceleration × time²)

C. final_velocity² = initial_velocity² + (2 × acceleration × distance_moved)

D. final_velocity = initial_velocity + (2 × acceleration × distance_moved)

MCQ 2: Kinetic friction is always

A. lesser than static friction

B. greater than static friction

C. equal to static friction

D. equal to contact force

MCQ 3: Gravitational potential is always

A. positive

B. negative

C. zero

D. infinity

MCQ 4: In order to keep a body moving in a circle, there exists a force on it that is directed toward the center of the circle. This force is known as

A. Centrifugal force

B. Centripetal force

C. Gravitational Force

D. magnetic force

MCQ 5: A rectangle-shaped open-to-sky tank of water has a length of 2 m and a width of 1 m. If the atmospheric pressure is assumed to be 100 kPa and thickness of the tank walls is assumed to be negligible, the force exerted by the atmosphere on the surface of water is

A. 20 kN

B. 50 kN

C. 100 kN

D. 200 kN

MCQ 6: If we have a positive and a negative charge, then force between them is

A. positive

B. negative

C. zero

D. infinite

MCQ 7: Electrical force exerted by two point charges on each other is inversely proportional to

A. sum of their charges

B. product of their charges

C. distance between them

D. square of distance between them

MCQ 8: Unit for pressure used in weather maps is millibar. 1 millibar is equal to

A. 1000 bar

B. 100 kPa

C. 100 Pa

D. 1 atm

MCQ 9: Speed of stationary waves is

A. 1 ms^{-1}

B. 2 ms^{-1}

C. 3 ms^{-1}

D. zero

MCQ 10: If charge is placed at infinity, its potential is

A. zero

B. infinite

C. 1

D. -1

MCQ 11: Most appropriate instrument for measurement of internal and external diameter of a tube is

A. vernier callipers

B. micrometer screw gauge

C. meter rule

D. measuring tape

MCQ 12: When distance from center is doubled then electric field strength will

A. decrease by the factor of four

B. increase by factor of four

C. will be same

D. decrease by factor of two

MCQ 13: Liquid A and liquid B exert same amount of pressure on each other, but the density of A is twice the density of B. The height of liquid B is 10 cm, then the height of liquid A would be

A. 5 cm

B. 10 cm

C. 20 cm

D. 40 cm

MCQ 14: Incorrect statement for co-efficient of friction could be that

A. The coefficient of friction is denoted by the Greek letter μ.

B. The coefficient of friction is directly proportional to the force of friction

C. The coefficient of friction is constant even in the conditions of fast slipping and high contact pressure

D. The coefficient of friction is inversely proportional to the force pressing the surfaces together

MCQ 15: If we move a positive charge to a positive plate, then potential energy of charge is

A. decreased

B. increased

C. remains constant

D. dissipated

MCQ 16: An instrument commonly used for the measurement of atmospheric pressure is known as

A. Manometer

B. Barometer

C. Calorimeter

D. Potentiometer

MCQ 17: Phase difference between a node and an antinode is

A. 90°

B. 45°

C. 180°

D. 360°

MCQ 18: Our weight, as measured by the spring weighing machine is equivalent of

A. The total gravitational force that Earth exerts on us

B. The total centripetal force required to keep us moving on Earth's axis

C. The total gravitational force that Earth exerts on us + The total centripetal force required to keep us moving on Earth's axis

D. The total gravitational force that Earth exerts on us - The total centripetal force required to keep us moving on Earth's axis

MCQ 19: Point where spring oscillates with maximum amplitude is called

A. node

B. antinode

C. fixed end

D. movable end

MCQ 20: According to Newton's law of universal gravitation, any two particles of finite mass attract one another with a force which is

A. Inversely proportional to the product of their masses and directly proportional to the square of their distance apart

B. Inversely proportional to the product of their masses and directly proportional to their distance apart

C. Directly proportional to the product of their masses and inversely proportional to their distance apart

D. Directly proportional to the product of their masses and inversely proportional to the square of their distance apart

MCQ 21: Correct example of vector quantities could be

A. Distance and Speed

B. Displacement and Velocity

C. Distance and Displacement

D. Speed and Velocity

MCQ 22: In a stationary wave, nodes are at

A. fixed points

B. movable points

C. there are no nodes

D. random points

MCQ 23: In the formation of stationary waves, at T/2 the waves are

A. out of phase

B. in phase

C. diminished

D. twice the amplitude

MCQ 24: A vector quantity is one which has

A. direction as well as magnitude

B. magnitude only

C. direction only

D. no direction, no magnitude

MCQ 25: Centripetal force is dependent on three factors, the odd one out of these factors is

A. Mass of the rotating object

B. Speed of the rotating object

C. Volume of the rotating object

D. Path radius

MCQ 26: Graph of potential energy against distance is

A. curve

B. parabolic

C. hyperbolic

D. straight line

MCQ 27: On all instruments like measuring cylinder, pipette and burette, readings are always taken at the bottom of the meniscus of liquid. This is not true for liquids like

A. Oil

B. Ink

C. Mercury

D. Seawater

MCQ 28: In a weather map, lines joining all those regions with same atmospheric pressure are called

A. Bars

B. Millibars

C. Isobars

D. Atmospheric bars

MCQ 29: For a given system, the minimum frequency of a standing wave is in a

A. fundamental mode

B. lowest mode

C. highest mode

D. peak mode

MCQ 30: In an electric field, energy per unit positive charge is

A. voltage

B. current

C. frequency

D. resistance

MCQ 31: In Kundt's dust tube, dust accumulates at

A. nodes

B. antinodes

C. at the end

D. at troughs only

MCQ 32: A node having twice the frequency of the fundamental is called

A. half harmonic

B. harmonic

C. double harmonic

D. triple harmonic

MCQ 33: If frequency of certain wave is f, then its velocity is

A. $v = f\lambda$

B. $v = T\lambda$

C. T^2

D. $1/_{T^2}$

MCQ 34: Origin of gravitational field is

A. charges

B. masses

C. Earth's core

D. matter

MCQ 35: At separation between a node and antinode, wavelength becomes

A. λ

B. $\lambda/_2$

C. $\lambda/4$

D. 2λ

Chapter 4

Capacitance MCQs

MCQ 1: If the plates of capacitor are oppositely charged then the total charge is equal to

A. negative

B. positive

C. zero

D. infinite

MCQ 2: Area under current-time graph represents

A. magnitude of charge

B. dielectric

C. amount of positive charge

D. amount of negative charge

MCQ 3: If charge stored on plates of capacitor is large, then capacitance will be

A. small

B. large

C. zero

D. infinite

MCQ 4: insulator which is placed between 2 plates of capacitor is

A. electric

B. dielectric

C. inductor

D. resistor

MCQ 5: Combined capacitance is equal to the

A. sum of all capacitance of capacitors

B. product of all the capacitance

C. difference between the capacitors

D. average capacitance of capacitors

MCQ 6: capacitance and charge on plates are

A. inversely related

B. directly related

C. not related at all

D. always equal

MCQ 7: If the capacitors are connected in parallel, then the potential difference across each capacitor is

A. same

B. different

C. zero

D. infinite

MCQ 8: Total capacitance of 300 mF capacitor and a 600 mF in series is

A. 300 mF

B. 500 mF

C. 200 mF

D. 1000 mF

MCQ 9: Work done in charging a capacitor is given by

A. $(1/2)QV$

B. $2QV$

C. QV

D. $2V$

MCQ 10: Device used to store energy in electrical circuits is

A. resistor

B. inductor

C. capacitor

D. diode

MCQ 11: Energy stored in a 2000 mF capacitor charged to a potential difference of 10 V is

A. 0.12 J

B. 1.3 J

C. 0.10 J

D. 3 J

MCQ 12: Capacitor is fully charged if potential difference is equal to

A. EMF

B. current

C. resistance

D. power

Chapter 5

Charged Particles MCQs

MCQ 1: An electron is travelling at right angles to a uniform magnetic field of flux density 1.2 mT with a speed of 8×10^6 ms^{-1}, the radius of circular path followed by electron is

A. 3.8 cm

B. 3.7 cm

C. 3.6 cm

D. 3.5 cm

MCQ 2: Hall voltage is directly proportional to

A. current

B. electric field

C. magnetic flux density

D. all of above

MCQ 3: Force due to magnetic field and velocity is

A. at right angles to each other

B. at acute angles with each other

C. at obtuse angle with each other

D. antiparallel to each other

MCQ 4: Force on a moving charge in a uniform magnetic field depends upon

A. magnetic flux density

B. the charge on the particle

C. the speed of particle

D. all of above

MCQ 5: Electric field strength related to hall voltage is given by

A. $V_H d$

B. V_H/d

C. $V_H E$

D. Ed

MCQ 6: Hall probe is made up of

A. metals

B. non metals

C. semiconductor

D. radioactive material

MCQ 7: For an electron, magnitude of force on it is

A. BeV

B. eV

C. Be

D. BV

MCQ 8: When current is parallel to magnetic fields, force on conductor is

A. zero

B. infinite

C. 2 times

D. same

MCQ 9: Direction of conventional current is

A. direction of neutron flow

B. direction of electron flow

C. direction of flow of positive charge

D. same as that of electric current

MCQ 10: According to the equation 'r = (mv)/(Be)', the faster moving particles

A. move in smaller circle

B. move straight

C. move in bigger circle

D. move randomly

MCQ 11: In Hall Effect, voltage across probe is known as

A. hall voltage

B. EMF

C. potential difference

D. hall potential

Chapter 6

Circular Motion MCQs

MCQ 1: Angle through which an object has moved is called it's

A. linear displacement

B. linear distance

C. angular displacement

D. angular distance

MCQ 2: Angular velocity of second hand of clock is 0.105 rad s^{-1} and length of hand is 1.8 cm, then speed of tip of hand is

A. 0.189 cms^{-1}

B. 1 cms^{-1}

C. 0.189 ms^{-1}

D. 2 ms^{-1}

MCQ 3: Object moving along a circular path is

A. in equilibrium

B. not in equilibrium

C. not moving with constant speed

D. in random motion

MCQ 4: At the fairground, the force that balances out our weight is

A. centripetal force

B. centrifugal force

C. friction

D. gravitational force

MCQ 5: If an object moves a circular distance 's' of radius 'r', then it's angular displacement is

A. s/r

B. r/s

C. rs

D. r^2s

MCQ 6: Speed of an object travelling around a circle depends upon

A. angular velocity

B. radius

C. circumference

D. both A and B

MCQ 7: Number of degree a complete circle represents is

A. 90°

B. 180°

C. 270°

D. 360°

MCQ 8: When a body is moving along a circular path, it's velocity is directed towards

A. center

B. normal

C. tangent

D. parallel to circle

MCQ 9: Speed is unchanged because work done on an object is

A. zero

B. positive

C. negative

D. infinite

MCQ 10: 180° is equal to

A. 2π rad

B. π rad

C. $\pi/2$ rad

D. $\pi/4$ rad

MCQ 11: For the minute hand of the clock, the angular velocity is equal to

A. 2 rad s^{-1}

B. 3 rad s^{-1}

C. 1 rad s^{-1}

D. 0.00175 rad s^{-1}

MCQ 12: 105° in radians is equal to

A. 2 rad

B. 3 rad

C. 1.83 rad

D. 4.5 rad

MCQ 13: 1 rad is equal to

A. 57.3°

B. 54°

C. 45°

D. 90°

MCQ 14: According to Newton's 2nd law the object's acceleration and centripetal force are

A. at right angles to each other

B. anti-parallel to each other

C. make acute angle with each other

D. in same direction

MCQ 15: A stone whirling in a horizontal circle on the end of string depicts

A. conical pendulum

B. cone

C. pendulum

D. eclipse

MCQ 16: Centripetal force is directed towards the

A. tangent to circle

B. center

C. normal to circle

D. parallel to circle

MCQ 17: Velocity required by an object to orbit around Earth is

A. 9 kms^{-1}

B. 7 kms^{-1}

C. 8 kms^{-1}

D. 10 kms^{-1}

Chapter 7

Communication Systems MCQs

MCQ 1: As compared to sound waves the frequency of radio waves is

A. lower

B. higher

C. equal

D. may be higher or lower

MCQ 2: Decrease in strength of signal is known as

A. tuning

B. modulation

C. attenuation

D. amplification

MCQ 3: If frequency of modulated wave is less than frequency of carrier wave, then input signal is

A. negative

B. positive

C. zero

D. infinite

MCQ 4: At the end of communication system, the signal is converted from radio to

A. sound

B. mechanical energy

C. kinetic energy

D. potential energy

MCQ 5: Energy is lost in wires due to

A. heating

B. resistance

C. conduction

D. both A and B

MCQ 6: Voltage signal generated by a microphone is

A. digital in nature

B. analogue in nature

C. consists of bits and bytes

D. hybrid in nature

MCQ 7: Phenomena in which signal transmitted in one circuit creates undesired effect in other circuit is known as

A. crosstalk

B. signal attenuation

C. sampling

D. crosslinking

MCQ 8: Digital number 9 can be represented in binary number as

A. 110

B. 1001

C. 1010

D. 1011

MCQ 9: Geostationary satellite has period

A. twice of Earth

B. same as Earth

C. half of Earth

D. quarter of Earth

MCQ 10: Bit on left hand side has

A. lowest value

B. zero value

C. highest value

D. infinite value

MCQ 11: A digital quantity has

A. only 2 values

B. more than 2 values

C. no values

D. less than 2 values

MCQ 12: Frequency of sky waves is

A. less than 3 MHz

B. more than 3 MHz

C. less than 2 MHz

D. exactly 2 MHz

MCQ 13: Each digit in a binary number is known as a

A. bit

B. byte

C. number

D. digit

MCQ 14: Value of sampled signal is used to produce a

A. binary number

B. decimal number

C. octal number

D. all of above

MCQ 15: Data in compact disc is stored in form of

A. analogue signal

B. digital signal

C. noise

D. colors

MCQ 16: With a 30 m long coaxial cable, the bandwidth can exceed

A. 100 MHz

B. 1000 MHz

C. 50 MHz

D. 300 MHz

MCQ 17: Amplitude of modulated wave is in phase with

A. output

B. system

C. frequency

D. signal

MCQ 18: Variation in amplitude or frequency of carrier wave is called

A. amplitude modulation

B. frequency modulation

C. modulation

D. bandwidth

MCQ 19: In frequency modulation, amplitude of modulated wave is

A. positive

B. negative

C. constant

D. zero

MCQ 20: High quality music only needs frequencies up to

A. 10 Hz

B. 15 Hz

C. 20 kHz

D. 15 kHz

MCQ 21: Binary system has base

A. 10

B. 11

C. 1

D. 2

MCQ 22: In FM, frequency of modulated wave varies with

A. amplitude

B. time

C. wavelength

D. energy

MCQ 23: Unwanted signal that distorts a transmitted signal is called

A. analogue

B. noise

C. digital

D. tuning

MCQ 24: First communication satellites used frequencies of

A. 6 GHz for uplink

B. 3 GHz for uplink

C. 6 GHz for downlink

D. 5 GHz for downlink

MCQ 25: A wave of frequency 1 GHz has wavelength of

A. 0.4 m

B. 0.5 m

C. 0.2 m

D. 0.3 m

Chapter 8

Electric Current, Potential Difference and Resistance MCQs

MCQ 1: Semiconductors have electron number density of order

A. 10^{24} m^{-3}

B. 10^{20} m^{-3}

C. 10^{12} m^{-3}

D. 10^{23} m^{-3}

MCQ 2: A straight line symbol shows the

A. fuse

B. diode

C. connecting lead

D. switch

MCQ 3: Rate of flow of electric charge is

A. electric current

B. conventional current only

C. electronic current only

D. potential difference

MCQ 4: instrument which transfers energy to electric charges in a circuit is

A. battery

B. voltmeter

C. ammeter

D. galvanometer

MCQ 5: Electric power is related to

A. current in component

B. potential difference

C. electrical resistance

D. both A and B

MCQ 6: If a current of 1 A passes through a point in 1 s then charge passes that point is

A. 2 C

B. 3 C

C. 1 C

D. 6 C

MCQ 7: Energy transferred per unit charge is

A. EMF

B. current

C. potential difference

D. conventional current

MCQ 8: Current in a circuit when a charge of 180 C passes a point in a circuit in 2 minutes is

A. 1:00 AM

B. 2:00 AM

C. 3:00 AM

D. 1.5 A

MCQ 9: Electrons which are free to move around are also called

A. conduction electrons

B. valence shell electrons

C. inner electrons

D. electron sea

MCQ 10: Number of conduction electrons per unit volume is

A. electron number

B. number density

C. proton number

D. neutron number

MCQ 11: Current in a circuit if resistance of 15 Ω and potential difference of 3.0 V is applied across its ends is

A. 1:00 AM

B. 2:00 AM

C. 0.5 A

D. 0.20 A

MCQ 12: Number density for copper is

A. 10^{-29} m^{-3}

B. 10^{30} m^{-3}

C. 10^{29} m^{-3}

D. 10^{20} m^{-3}

MCQ 13: Mean drift velocity of electron in a copper wire having cross-sectional area 5.0×10^{-6} m² carrying current of 1 A and having number density 8.5×10^{28} m³ is

A. 0.015 mms^{-1}

B. 0.1 mms^{-1}

C. 0.5 mms^{-1}

D. 0.25 mms^{-1}

MCQ 14: Magnitude of charge is known as

A. charge count

B. elementary charge

C. elementary count

D. charge number

MCQ 15: 1 Ω is equal to

A. 1 V A^{-2}

B. 1 V A^{-1}

C. 1 V^{-1} A

D. 2 V A^{-1}

MCQ 16: To protect wiring from excessive passing of current is

A. voltmeter

B. fuse

C. galvanometer

D. resistance

MCQ 17: Current in a 60 W light bulb when it is connected to a 230 V power supply is

A. 0.26 A

B. 1.5 A

C. 2.6 A

D. 3.9 A

MCQ 18: Grid cables are 15 km long with a resistance of 0.20 Ω km^{-1}, powers wasted as heat in these cables are

A. 50 kW

B. 60 kW

C. 20 kW

D. 30 kW

MCQ 19: Actual velocity of electrons between collisions is

A. 10^{30} ms^{-1}

B. 10^{20} ms^{-1}

C. 10^{2} ms^{-1}

D. 10^{5} ms^{-1}

MCQ 20: By increasing the current, the drift velocity

A. decreases

B. increases

C. remains constant

D. becomes zero

MCQ 21: Current in a circuit depends on

A. resistance

B. potential difference

C. both A and B

D. EMF

MCQ 22: If direction of current is from positive to negative, then it is called

A. electronic current

B. conventional current only

C. positronic current

D. protonic current

MCQ 23: A voltmeter arranged across the power supply measures

A. potential difference

B. EMF

C. current

D. resistance

Chapter 9

Electric Field MCQs

MCQ 1: Particles involved in the movement within material are

A. protons

B. electrons

C. neutrons

D. positrons

MCQ 2: Phenomena in which a charged body attract uncharged body is called

A. electrostatic induction

B. electric current

C. charge movement

D. magnetic induction

MCQ 3: An uncharged object has

A. more protons

B. more electrons

C. equal electrons and protons

D. no protons and electrons

MCQ 4: Fields that act on objects with masses are

A. electric fields

B. magnetic fields

C. force fields

D. gravitational fields

MCQ 5: Where an electric charge experiences a force, there is an

A. electric field

B. magnetic field

C. electric current

D. conventional current

MCQ 6: A field that spreads outwards in all directions is

A. radial

B. non radial

C. strong

D. weak

MCQ 7: At all the points the uniform fields have

A. different strength

B. same strength

C. zero strength

D. infinite strength

MCQ 8: Electric field strength on a dust particle having charge equal to 8×10^{-19} when plates are separated by distance of 2 cm and have a potential difference of 5 kV is

A. 2.0×10^{-13} N

B. 3 N

C. 5 N

D. 20 N

MCQ 9: Electric field strength can be defined as

A. $E = Q/F$

B. $E = W/F$

C. $E = F/Q$

D. $E = P/Q$

MCQ 10: When one material is rubbed against the other, then it becomes electrically

A. neutral

B. charged

C. positively charged

D. negatively charged

MCQ 11: When an electron is moving horizontally between oppositely charged plates, it will move in the

A. straight line

B. fall directly downwards

C. move towards positive plates

D. curved path

Chapter 10

Electromagnetic Induction MCQs

MCQ 1: EMF can be induced in a circuit by

A. changing magnetic flux density

B. changing area of circuit

C. changing the angle

D. all of above

MCQ 2: A straight wire of length 0.20 m moves at a steady speed of 3.0 ms^{-1} at right angles to the magnetic field of flux density 0.10 T. The EMF induced across ends of wire is

A. 0.5 V

B. 0.06 V

C. 0.05 V

D. 0.04 V

MCQ 3: By accelerating the magnet inside the coil, the current in it

A. increases

B. decreases

C. remains constant

D. reverses

MCQ 4: Consequence of motor effect is

A. electromagnetic induction

B. current

C. voltage

D. EMF

MCQ 5: Total number of magnetic field lines passing through an area is called

A. magnetic flux density

B. magnetic flux

C. EMF

D. voltage

MCQ 6: Magnitude of induced EMF is proportional to

A. rate of change of current

B. rate of change of voltage

C. rate of change of magnetic flux linkage

D. rate of change of resistance

MCQ 7: In transformer, the core is made up of soft iron in order to pass the maximum

A. flux

B. current

C. magnetic flux

D. voltage

MCQ 8: For a straight wire, induced current depends upon

A. the speed of movement of wire

B. the length of wire

C. the magnitude of magnetic flux density

D. all of above

MCQ 9: In generators, the rate of change of flux linkage is maximum when the coil is moving through the

A. vertical position

B. horizontal position

C. diagonal position

D. at an angle of 45°

MCQ 10: EMF for a coil depends upon

A. the cross sectional area

B. no. of turns of wire

C. the magnitude of magnetic flux density

D. all of above

MCQ 11: Currents that flow in circles inside a disc are known as

A. eddy currents

B. circular currents

C. air currents

D. alternating curents

MCQ 12: When field is parallel to plane of area, magnetic flux through coil is

A. zero

B. infinite

C. 2

D. 5

MCQ 13: Moving a coil in and out of magnetic field induces

A. force

B. potential difference

C. EMF

D. voltage

MCQ 14: Induced current in coil by a magnet turns it into an

A. straight wire

B. magnet

C. ammeter

D. electromagnet

Chapter 11

Electromagnetism and Magnetic Field MCQs

MCQ 1: Strength of magnetic field is known as

A. flux

B. density

C. magnetic strength

D. magnetic flux density

MCQ 2: Unit of luminous intensity is

A. m

B. kg

C. cd

D. mol

MCQ 3: For a hydrogen atom the electrical force as compared to gravitational force is

A. 10^{39} times

B. 10^{40} times

C. 10^{41} times

D. 10^{42} times

MCQ 4: Weakest force in nature is

A. electric force

B. gravitational force

C. weak force

D. magnetic force

MCQ 5: Magnetic field can be produced by using

A. permanent magnet

B. electric current

C. temporary magnet

D. both A and B

MCQ 6: Whenever there is force on magnetic pole, there exists

A. magnetic field

B. electric field

C. current

D. voltage

MCQ 7: If the magnetic flux density and current are at right angles, then component of force acting on the conductor is

A. BIL $\cos\theta$

B. BIL $\sin\theta$

C. BIL $\tan\theta$

D. BL sinθ

MCQ 8: Unraveling an electromagnetic gives

A. stronger field

B. weaker field

C. moderate field

D. wider field

MCQ 9: Force per meter on two wires carrying a current of 1 A placed 1 m apart is equal to

A. 6.7×10^{-11} N

B. 9.0×10^{9} N

C. 2.0×10^{-7} N

D. 3.0×10^{-4} N

MCQ 10: F = BIL can only be used if the magnetic field and electric current are

A. at right angles to each other

B. in same direction

C. anti-parallel to each other

D. anti-perpendicular to each other

MCQ 11: Derived unit Tesla is related to

A. A

B. kg

C. s

D. all of above

MCQ 12: If we reverse the direction of electric current, then the direction of magnetic field will be

A. reversed

B. remains same

C. becomes tangent

D. becomes normal

MCQ 13: 1 Tesla is equal to

A. $50 \text{ N A}^{-1} \text{ m}^{-1}$

B. $100 \text{ N A}^{-1} \text{ m}^{-1}$

C. $1 \text{ N A}^{-1} \text{ m}^{-1}$

D. $1000 \text{ N A}^{-1} \text{ m}^{-1}$

MCQ 14: Field which does not have magnetic poles is

A. straight lined

B. normal to the wire

C. tangent to the wire

D. circular

MCQ 15: A flat coil and solenoid has

A. different fields

B. same physical properties

C. same fields

D. same chemical properties

MCQ 16: Flux density is defined by

A. FIL

B. F/(IL)

C. IL

D. I/FL

MCQ 17: Strength of magnetic field of solenoid can be increased by adding core made of

A. copper

B. ferrous

C. silver

D. aluminum

MCQ 18: In Fleming's left hand rule, thumb shows direction of

A. current

B. field

C. motion

D. charge

MCQ 19: If the Earth's magnetic field lines pass through the Hall probe in opposite direction, then the change in reading of voltmeter is

A. thrice the Earth's magnetic flux density

B. four times the Earth's magnetic flux density

C. five times the Earth's magnetic flux density

D. twice the Earth's magnetic flux density

Chapter 12

Electronics MCQs

MCQ 1: LED starts to conduct when voltage is about

A. 1 V

B. 4 V

C. 3 V

D. 2 V

MCQ 2: For non-inverting amplifier the input and output is

A. out of phase

B. in phase

C. have phase difference of 180°

D. have phase difference of 90°

MCQ 3: A sensing device is also called

A. transistor

B. thermistor

C. sensor

D. transducer

MCQ 4: Op-amp can provide maximum output current of

A. 25 mA

B. 30 mA

C. 35 mA

D. 40 mA

MCQ 5: Output resistance of an actual op-amp is

A. 45 Ω

B. 46 Ω

C. 70 Ω

D. 75 Ω

MCQ 6: Impedance of ideal op-amp is

A. zero

B. 1

C. infinite

D. 2

MCQ 7: Change in length and cross-sectional area of metal wire changes

A. current

B. voltage

C. resistance

D. magnetic effect

MCQ 8: Amplifier produces output with more

A. power only

B. voltage only

C. current only

D. power and voltage

MCQ 9: Number of power supplies required to get output of op-amp is

A. two

B. four

C. six

D. three

MCQ 10: A device used to avoid the relay destroying the op-amp is

A. diode

B. LED

C. reverse bias diode

D. forward biased diode

MCQ 11: As long as op-amp is not saturated, the potential difference between inverting and non-inverting inputs is

A. zero

B. infinite

C. 1

D. 2

MCQ 12: A light dependent resistor is made up of

A. low resistance semiconductor

B. low resistance metal

C. high resistance semiconductor

D. high resistance metal

MCQ 13: In inverting amplifier, the phase difference between input and output voltages must be

A. 30°

B. 45°

C. 90°

D. 180°

MCQ 14: An infinite slew rate refers to

A. no time delay

B. small time delay

C. large time delay

D. variable time delay

MCQ 15: To limit current in LED, resistor should be connected in

A. parallel

B. series

C. evacuated flask

D. ionized tube

MCQ 16: A component whose property changes when there is a change in any physical quantity of a device is

A. processor

B. sensor

C. output device

D. portable device

MCQ 17: Benefits of negative feedback include

A. less distortion

B. increased bandwidth

C. low output resistance

D. all of above

MCQ 18: When temperature rises, resistance of negative temperature coefficient thermistor

A. increases

B. decreases

C. zero

D. infinity

MCQ 19: If current of 20 mA is passing through op-amp and voltage drop across series resistor is 10 V then value of resistance is

A. 500Ω

B. 400Ω

C. 300Ω

D. 200Ω

MCQ 20: Graph of resistance of thermistor to temperature is

A. exponential decrease

B. linear decrease

C. exponential increase

D. linear increase

MCQ 21: Output current from an LED is

A. 20 mA

B. 30 mA

C. 40 mA

D. 50 mA

MCQ 22: In an inverting amplifier the non-inverting input (+) is connected to

A. 1 V line

B. 0 V line

C. 2 V line

D. 3 V line

MCQ 23: Actual op-amp may have an open loop gain of

A. 10^2

B. 10^5

C. 10^3

D. 10^4

MCQ 24: Coil of relay is connected to the

A. input of op-amp

B. output of op-amp

C. midpoint of op-amp

D. anywhere in between input and output of op-amp

Chapter 13

Forces, Vectors and Moments MCQs

MCQ 1: Pair of forces that cause steering wheel of a car to rotate is called

A. couple

B. friction

C. normal force

D. weight

MCQ 2: If the principle of moments for any object holds, then object is in state of

A. inertia

B. equilibrium

C. suspension

D. motion

MCQ 3: Combined effect of several forces is known as

A. net force

B. resultant force

C. normal force

D. weight

MCQ 4: To form a couple, the force should be

A. equal in magnitude

B. parallel and opposite

C. separated by distance

D. all of above

MCQ 5: Moment of force depends upon

A. magnitude of force

B. perpendicular distance of force from pivot

C. both A and B

D. axis of rotation

MCQ 6: Two perpendicular components are

A. independent of each other

B. dependent on each other

C. anti-parallel to each other

D. parallel to each other

MCQ 7: Object is in equilibrium if resultant force acting on it is

A. increasing

B. decreasing

C. zero

D. becomes constant

MCQ 8: Center of gravity of an irregular body lies on the

A. edge

B. center of body

C. point of intersection of lines

D. along the axis of rotation

MCQ 9: Number of forces a falling tennis ball experiences is

A. 1

B. 3

C. 2

D. 4

MCQ 10: Point where all weight of object acts is called

A. central point

B. center of gravity

C. edge

D. center of mass

MCQ 11: If weight of a falling tennis ball is 1.0 N and drag force acting on it is 0.2 N then resultant force is

A. 0.8 N

B. 0.5 N

C. 1 N

D. 2 N

MCQ 12: If the resultant vector forms an angle of 45°, then the two components are

A. parallel to each other

B. perpendicular to each other

C. anti-parallel to each other

D. anti-perpendicular to each other

Chapter 14

Gravitational Field MCQs

MCQ 1: Force acting on two point masses is directly proportional to

A. sum of masses

B. difference of masses

C. distance between masses

D. product of masses

MCQ 2: Mass of Earth when it's radius is 6400 km and gravitational field strength is 9.81 N kg^{-1} is

A. 6.0×10^{24} kg

B. 5×10^{23} kg

C. 40×10^{9} kg

D. 9×10^{24} kg

MCQ 3: On scale of building, gravitational field is

A. increasing

B. decreasing

C. uniform

D. varying

MCQ 4: Decrease in field strength on top of Mount Everest is

A. 10%

B. 5%

C. 1%

D. 0.30%

MCQ 5: Work done on an object to bring it to certain point in space is called

A. gravitational potential energy

B. potential energy

C. kinetic energy

D. mechanical energy

MCQ 6: N kg^{-1} is equivalent to

A. ms^{-2}

B. ms^{-1}

C. ms

D. ms^{-3}

MCQ 7: Mass of satellite orbiting Earth is

A. considered

B. irrelevant

C. should be infinite

D. should be zero

MCQ 8: Spacing between field lines indicates

A. direction of field lines

B. strength of field lines

C. magnitude of field lines

D. work done by field lines

MCQ 9: Square of orbital period is proportional to

A. radius

B. square of radius

C. cube of radius

D. square of diameter

MCQ 10: Number of satellites in geostationary orbits are

A. 100

B. 200

C. 300

D. 400

MCQ 11: Geostationary satellites have lifetime of nearly

A. 20 years

B. 10 years

C. 50 years

D. 60 years

MCQ 12: Time taken to complete a revolution around a planet is called

A. orbital period

B. time period

C. frequency

D. wavelength

MCQ 13: As distance increases, value of gravitational field strength

A. also increases

B. decreases

C. remains constant

D. may increase or decrease

MCQ 14: All objects are attracted towards

A. center of Earth

B. sun

C. mars

D. moon

MCQ 15: Satellite around Earth follows a circular path because

A. gravitational force is parallel to velocity

B. gravitational force is anti-parallel to velocity

C. gravitational force is perpendicular to velocity

D. gravitational force is anti-perpendicular to

MCQ 16: Each 1 kg mass experiences force of

A. 7 N

B. 9.81 N

C. 20 N

D. 100 N

MCQ 17: Gravitational potential is always

A. infinite

B. zero

C. positive

D. negative

MCQ 18: The closer satellite is to Earth, its speed should be

A. more fast

B. more slow

C. zero

D. any constant value

Chapter 15

Ideal Gas MCQs

MCQ 1: Normal force exerted per unit area by gas on walls of container is

A. temperature

B. energy

C. pressure

D. friction

MCQ 2: Surface area of a typical person is about

A. 1 m²

B. 2 m²

C. 3 m²

D. 4 m²

MCQ 3: If we double the temperature of an ideal gas, then its average kinetic energy will be

A. halved

B. triple the original

C. fourth times of original

D. doubled

MCQ 4: Escape velocity for a particle is about

A. 5 kms^{-1}

B. 8 kms^{-1}

C. 11 kms^{-1}

D. 14 kms^{-1}

MCQ 5: Force exerted on a person by atmosphere is

A. 200 000 N

B. 300 000 N

C. 400 000 N

D. 500 000 N

MCQ 6: 10 mole of carbon contains

A. 6.02×10^{24} atoms

B. 6.03×10^{23} atoms

C. 6.02×10^{23} atoms

D. 6.04×10^{24} atoms

MCQ 7: Space occupied by gas is

A. area

B. volume

C. space

D. mass

MCQ 8: At absolute zero, volume of gas is equal to

A. $0 \ m^3$

B. $1 \ m^3$

C. $2 \ m^3$

D. $3 \ m^3$

MCQ 9: Number of moles in 1.6 kg of oxygen is

A. 30 mol

B. 50 mol

C. 40 mol

D. 60 mol

MCQ 10: Quantity R/N_A defines

A. Plank's constant

B. Boltzmann constant

C. gravitational constant

D. Avogadro's constant

MCQ 11: Molar gas constant has value

A. $7 \ J \ mol^{-1} \ K^{-1}$

B. $8 \ J \ mol^{-1} \ K^{-1}$

C. $8.31 \ J \ mol^{-1} \ K^{-1}$

D. 5 J mol^{-1} K^{-1}

MCQ 12: If average kinetic energy of molecules is higher, then temperature of gas is

A. high

B. low

C. zero

D. infinite

MCQ 13: The ideal gas equation is

A. PV = constant

B. PT = constant

C. V/T

D. PV = nRT

MCQ 14: Graph of p against 1/V is

A. curve

B. straight line

C. parabola

D. hyperbola

MCQ 15: Gases deviate from gas laws at

A. high temperature only

B. low pressure only

C. high pressure and low temperature

D. low temperature only

MCQ 16: Law which relates pressure and volume of gas is

A. Charles's law

B. Avogadro's law

C. Boyle's law

D. ideal gas law

MCQ 17: As compared to the volume occupied by gas, the volume of particles is

A. more

B. infinite

C. negligible

D. less than the volume of gas

MCQ 18: Pressure of gas depends on the

A. density of gas

B. mean square speed of gas molecules

C. both A and B

D. temperature

MCQ 19: At standard pressure and temperature the average speed of molecules is

A. 400 ms^{-1}

B. 500 ms^{-1}

C. 600 ms^{-1}

D. 700 ms^{-1}

Chapter 16

Kinematics Motion MCQs

MCQ 1: Speed of sound in water is 1500 ms^{-1}, depth of water when reflected sound waves are detected after 0.40 s is

A. 700 m

B. 600 km

C. 600 m

D. 750 km

MCQ 2: If a snail crawls 12 cm in 60 s then its average speed in mms^{-1} is

A. 2 mms^{-1}

B. 5 mms^{-1}

C. 10 mms^{-1}

D. 20 mms^{-1}

MCQ 3: Speed of a body in particular direction can be called

A. acceleration

B. displacement

C. velocity

D. distance

MCQ 4: If distance is increasing uniformly with time, then velocity is

A. increasing

B. decreasing

C. constant

D. zero

MCQ 5: If the slope of a graph is zero and so the displacement, then velocity is

A. increasing

B. decreasing

C. constant

D. zero

MCQ 6: Speed of snail as compared to speed of a car will be in units

A. cms^{-1}

B. ms^{-1}

C. kms^{-1}

D. $km\ h^{-1}$

MCQ 7: If a river flows from west to east with constant velocity of 1.0 ms^{-1} and a boat leaves south bank heading towards north with velocity of 2.40 ms^{-1}, then resultant velocity of boat is

A. 2.6 ms^{-1}

B. 2.7 ms^{-1}

C. 2.8 ms^{-1}

D. 2.9 ms^{-1}

MCQ 8: Two sides of a rectangular table are 0.8 m and 1.2 m, the displacement of spider when it runs a distance of 2.0 m is

A. 1.5 m

B. 1.4 m

C. 1.6 m

D. 1.7 m

MCQ 9: Distance travelled by a body in time 't' is

A. instantaneous speed

B. average velocity

C. average acceleration

D. instantaneous acceleration

MCQ 10: Displacement is a

A. scalar quantity

B. vector quantity

C. base quantity

D. derived quantity

MCQ 11: A car is travelling at 15 ms^{-1}, the distance it will travel in 1 hour is

A. 54 km

B. 55 km

C. 56 km

D. 57 km

MCQ 12: A car travelled south-west for 200 miles depicts

A. distance

B. speed

C. velocity

D. displacement

Chapter 17

Kirchhoff's Laws MCQs

MCQ 1: Sum of EMF around any loop equals

A. sum of current

B. sum of potential difference

C. sum of resistances

D. sum of charges passing through it per second

MCQ 2: Current entering and leaving a point in a circuit should be

A. equal

B. decreasing

C. increasing

D. variable

MCQ 3: Current through each resistor when they are connected in series is

A. different

B. same

C. can be both A and B

D. decreasing

MCQ 4: If a billion electrons enter a point in 1 s, then number of electrons leaving that point in 1 s are

A. 2 billion

B. 3 billion

C. 1 billion

D. 10 billion

MCQ 5: Three resistances 20 Ω, 30 Ω and 60 Ω are connected in parallel, their combined resistance is given by

A. 110 Ω

B. 50 Ω

C. 20 Ω

D. 10 Ω

MCQ 6: Kirchhoff's 2nd law deals with

A. current in circuit

B. voltage in circuit

C. EMF in circuit

D. both B and C

MCQ 7: Combined resistance of 5 Ω and 10 Ω is equal to

A. 10 Ω

B. 16 Ω

C. 15 Ω

D. 20 Ω

MCQ 8: Two 10 Ω resistors are connected in parallel, their equivalent resistance is

A. 5 Ω

B. 0.2 Ω

C. 15 Ω

D. 20 Ω

MCQ 9: Kirchhoff's 2nd law is consequence of law of conservation of

A. energy

B. charge

C. momentum

D. power

MCQ 10: Ammeter should always have a

A. high resistance

B. low resistance

C. low voltage

D. high voltage

MCQ 11: 1 V is equal to

A. 1 J C^{-1}

B. 2 J C^{-1}

C. 1 J^{-1} C

D. 2 J⁻¹ C

MCQ 12: Ideal resistance of ammeter is

A. 1 Ω

B. 2 Ω

C. 0 Ω

D. infinite

Chapter 18

Matter and Materials MCQs

MCQ 1: Density of water in kg m^{-3} is

A. 1000

B. 100

C. 10 000

D. 4000

MCQ 2: Normal force acting per unit cross sectional area is called

A. weight

B. pressure

C. volume

D. friction

MCQ 3: Ratio of tensile to strain is

A. Young's modulus

B. stress

C. stiffness

D. tensile force

MCQ 4: Gradient of force-extension graph is

A. variable

B. increasing

C. decreasing

D. force constant

MCQ 5: Energy in deformed solid is called

A. stress energy

B. potential energy

C. kinetic energy

D. strain energy

MCQ 6: Units of stress are

A. Newton

B. Joules

C. Pascal

D. Watt

MCQ 7: Concentration of matter in a material is

A. volume

B. mass

C. density

D. weight

MCQ 8: Units of strain are

A. Newton

B. Joules

C. Watt

D. no units

MCQ 9: Stress is force applied on

A. volume

B. cross sectional area

C. unit length

D. across diagonal

MCQ 10: The larger the spring constant, the spring would be more

A. extensible

B. stiffer

C. compressive

D. brittle

MCQ 11: Height of atmosphere, if atmospheric density is 1.29 kg m^{-3} and atmospheric pressure is 101 kPa, is

A. 7839.4 m

B. 7829.4 m

C. 7849.4 m

D. 7859.4 m

MCQ 12: Pressure in fluid depends upon

A. depth below the surface

B. density of fluid

C. the value of g

D. all of above

MCQ 13: As depth increases, pressure in a fluid

A. increases

B. decreases

C. remains constant

D. varies

MCQ 14: If extension in spring is proportional to load applied, then material obeys

A. Newton's law

B. gravitational law

C. Charles's law

D. Hooke's law

MCQ 15: If a spring is squashed, then forces are

A. extensible

B. compressive

C. normal

D. abnormal

MCQ 16: Fractional increase in original length is called

A. stress

B. strain

C. tensile force

D. compression

MCQ 17: If a force of 50 N is applied across the cross-sectional area of $5 \times 10^{-7} m^2$ then stress applied on it is

A. 1×10^8 Pa

B. 20 Pa

C. 50 Pa

D. 100 Pa

MCQ 18: Mass of steel sphere having density 7850 kg m^{-3} and radius 0.15 m is

A. 112 kg

B. 290 kg

C. 110.9 kg

D. 300 kg

MCQ 19: In a force-extension graph, force is taken along horizontal axis because

A. force is independent variable

B. extension is independent variable

C. force is dependent variable

D. all of above

MCQ 20: Extension and applied force are

A. directly proportional

B. inversely proportional

C. are independent of each other

D. inversely related

MCQ 21: 1 Pa is equal to

A. 1N m

B. 1 kg

C. 1 N m^{-1}

D. 1 N m^{-2}

MCQ 22: Spring constant of spring is also called

A. gradient

B. tensile forces

C. stiffness

D. compression

Chapter 19

Mechanics and Properties of Matter MCQs

MCQ 1: Tensile strain is equal to

A. Force per unit area

B. Force per unit volume

C. Extension per unit length

D. Force per unit length

MCQ 2: In elastic collisions,

A. only the total momentum of the colliding objects is conserved.

B. only the total kinetic energy is conserved.

C. both of the momentum and total kinetic energy are conserved.

D. neither momentum of the colliding bodies nor the total kinetic energy are recoverable.

MCQ 3: Total angular momentum of a body is given by

A. $I \times \omega$; where I: moment of inertia of the body, ω: angular velocity

B. $I^2 \times \omega$; where I: moment of inertia of the body, ω: angular velocity

C. $I^2 \times \omega^2$; where I: moment of inertia of the body, ω: angular velocity

D. $I \times \omega^2$; where I: moment of inertia of the body, ω: angular velocity

MCQ 4: Force that acts on a mass of 1 g and gives it an acceleration of 1 cms^{-2} is defined as

A. 1 newton

B. 1 dyne

C. 1 pound-force

D. 1 pa-force

MCQ 5: An object moving in a circle of radius 'r' with a constant speed 'v' has a constant acceleration towards the center equal to

A. v^2/r

B. v/r

C. $v^2 \times r$

D. $v \times r$

MCQ 6: Einstein's mass-energy relationship states that if the mass decreases by Δm, the energy released ΔE is given by

A. $\Delta E = \Delta m \times c$, where "c" denotes the speed of light.

B. $\Delta E = \Delta m \times c^2$, where "c" denotes the speed of light.

C. $\Delta E = \Delta m / c$, where "c" denotes the speed of light.

D. $\Delta E = \Delta m / c^2$, where "c" denotes the speed of light.

MCQ 7: Bernoulli's principle states that, for streamline motion of an incompressible non-viscous fluid:

A. the pressure at any part + the kinetic energy per unit volume = constant

B. the kinetic energy per unit volume + the potential energy per unit volume = constant

C. the pressure at any part + the potential energy per unit volume = constant

D. the pressure at any part + the kinetic energy per unit volume + the potential energy per unit volume = constant

MCQ 8: While Young's modulus 'E' relates to change in length and bulk modulus 'K' relates to change in volume, modulus of rigidity 'G' relates to change in:

A. weight

B. density

C. shape

D. temperature

MCQ 9: Young's modulus is defined as

A. tensile strain/tensile stress

B. tensile stress/tensile strain

C. tensile stress × tensile strain

D. length/area

MCQ 10: Velocity of escape is equal to

A. $r \times \sqrt{(2g)}$; where r: radius of Earth or any other planet for that matter, g: gravitational field strength

B. $g \times \sqrt{(2r)}$; where r: radius of Earth or any other planet for that matter, g: gravitational field strength

C. $\sqrt{(2g)}/r$; where r: radius of Earth or any other planet for that matter, g: gravitational field strength

D. $\sqrt{(2gr)}$; where r: radius of Earth or any other planet for that matter, g: gravitational field strength

MCQ 11: Speed 'v' with which wave travels through a medium is given by

A. modulus of elasticity/density of the medium

B. modulus of elasticity/$\sqrt{}$(density of the medium)

C. $\sqrt{}$(modulus of elasticity/density of the medium)

D. v=d/t

MCQ 12: Hooke's law states that

A. the extension is proportional to the load when the elastic limit is not exceeded

B. the extension is inversely proportional to the load when the elastic limit is not exceeded

C. the extension is independent of the load when the elastic limit is not exceeded

D. load is dependent on extension

MCQ 13: Dimensions of strain are

A. [L]

B. $[M] [L]^{-1} [T]^{-2}$

C. $[L]^{-1}$

D. It's a dimensionless quantity

MCQ 14: Due to energy dissipation by viscous forces in air, if simple harmonic variations of a pendulum die away after some time, then oscillation is said to be:

A. undamped

B. free

C. damped

D. dependent

MCQ 15: At 'yield point' of a copper wire

A. the load hasn't exceeded the elastic limit yet; so, Hooke's law applies

B. the load has already exceeded the elastic limit and the material has become plastic

C. even the plastic stage has passed and the wire has snapped already

D. Like Brass and Bronze, Copper has no yield point

MCQ 16: Stationary waves are also called

A. static waves

B. standing waves

C. progressive waves

D. All of the above

MCQ 17: When the work done in moving a particle round a closed loop in a field is zero, the forces in the field are called

A. Zero forces

B. Non-Conservative forces

C. Conservative forces

D. Viscous forces

MCQ 18: Substances that elongate considerably and undergo plastic deformation before they break are known as

A. brittle substances

B. breakable substances

C. ductile substances

D. elastic substances

MCQ 19: 1 torr is equal to

A. 1 N/m²

B. 1 mm Hg

C. 1 bar

D. All of the above

MCQ 20: Velocity of sound waves through any material depends on

A. the material's density 'd' only

B. the material's density 'd' as well as its modulus of elasticity 'E'

C. the material's modulus of elasticity 'E' only

D. neither the material's density 'd' nor its modulus of elasticity 'E'

MCQ 21: Period of simple harmonic motion of a spiral spring or elastic thread is given by

A. T = 2π × (extension produced/gravitational field strength)

B. T = 2π × (extension produced/√(gravitational field strength))

C. T = 2π × (√(extension produced)/gravitational field strength)

D. $T = 2\pi \times \sqrt{(\text{extension produced/gravitational field strength})}$

MCQ 22: In order to slip one surface over another, maximum frictional force has to be overcome, this maximum frictional force between the two surfaces is also known as

A. kinetic frictional force

B. maximal frictional force

C. limiting frictional force

D. resisting force

MCQ 23: Van der Waals derived an expression for the 'pressure defect', if the observed pressure is denoted as 'p' and volume is denoted as 'V', the gas pressure in the bulk of the gas is equal to:

A. $p + a/V$; where a: constant for the particular gas

B. $p + a/(V^2)$; where a: constant for the particular gas

C. $p + (a \times V)$; where a: constant for the particular gas

D. $p + (a \times V^2)$; where a: constant for the particular gas

MCQ 24: "Upthrust = Weight of the liquid displaced" is known as

A. Bernoulli's Principle

B. Archimedes' Principle

C. Pascal's Law

D. Coulomb's law

MCQ 25: Assuming uniform density of the core, the acceleration due to gravity below the Earth's surface is

A. inversely proportional to the square of the distance from the center of the Earth

B. inversely proportional to the distance from the center of the Earth

C. directly proportional to the square of the distance from the center of the Earth

D. directly proportional to the distance from the center of the Earth

MCQ 26: When a gas or a liquid is subjected to an increased pressure, the substance contracts, the bulk strain is defined as

A. final volume / original volume

B. final pressure / original pressure

C. change in volume / original volume

D. original volume / change in volume

MCQ 27: Tensile stress is equal to

A. Force per unit area

B. Force per unit volume

C. Extension per unit length

D. Extension per unit area

MCQ 28: Dimensions of relative density are

A. mass $\times$ length^{-3}

B. mass $\times$ length^3

C. It has no dimensions, since it's a ratio of two densities

D. length^3 $\times$ mass^{-1}

MCQ 29: Dimensions of gravitational constant 'G' are:

A. $[M]^{-1} [L]^3 [T]^{-2}$

B. $[M] [L]^3 [T]^{-2}$

C. $[M]^{-1} [L]^2 [T]^{-1}$

D. $[M] [L]^{-1} [T]^2$

MCQ 30: A person of mass 'm' kg jumps from a height of 'h' meters, he will land on the ground with a velocity equal to:

A. $\sqrt{(2 \times g \times h)}$

B. $1/h \times \sqrt{(2 \times g)}$

C. $2gh$

D. $2\sqrt{(g \times h)}$

MCQ 31: In linear motion, the energy is given by $\frac{1}{2}mv^2$. Similarly, in rotational motion, the rotational energy is given by

A. $1/2 \times I \times \omega$; where I: moment of inertia of the body, ω: angular velocity

B. $1/2 \times I^2 \times \omega$; where I: moment of inertia of the body, ω: angular velocity

C. $1/2 \times I \times \omega^2$; where I: moment of inertia of the body, ω: angular velocity

D. $1/2 \times I^2 \times \omega^2$; where I: moment of inertia of the body, ω: angular velocity

MCQ 32: Boyle's law states that

A. pressure of a gas is inversely proportional to its volume i.e. $P \times V =$

constant

B. pressure of a gas is directly proportional to its volume i.e. $P/V =$ constant

C. pressure of a gas is inversely proportional to the square of its volume i.e. $P \times V^2 =$ constant

D. pressure of a gas is directly proportional to the square of its volume i.e. $P/V^2 =$ constant

MCQ 33: Isothermal bulk modulus is equal to

A. $Y \times P$; where Y: the ratio of the specific heat capacities of the gas, P: pressure

B. Pressure

C. The ratio of the specific heat capacities of the gas

D. Y/P; where Y: the ratio of the specific heat capacities of the gas, P: pressure

MCQ 34: Adiabatic bulk modulus is equal to:

A. $Y \times P$; where Y: the ratio of the specific heat capacities of the gas, P: pressure

B. Pressure

C. The ratio of the specific heat capacities of the gas

D. Y/P; where Y: the ratio of the specific heat capacities of the gas, P: pressure

MCQ 35: Bernoulli's principle shows that, at points in a moving fluid where the potential energy change is very small

A. the pressure is low where the velocity is low and similarly, the pressure is high where the velocity is high

B. the pressure is low where the velocity is high and conversely, the pressure is high where the velocity is low

C. pressure becomes independent of the velocity of the moving fluid

D. pressure remain independent of the speed of the stationary fluid

MCQ 36: 1 N (newton) is equal to

A. 10^2 dynes

B. 10^3 dynes

C. 10^4 dynes

D. 10^5 dynes

MCQ 37: Torricelli's theoremstates that the velocity 'v' of the liquid emerging from the bottom of the wide tank is given by $\sqrt{(2gh)}$. In practice, this velocity is:

A. equal to $\sqrt{(2gh)}$

B. greater than $\sqrt{(2gh)}$

C. lesser than $\sqrt{(2gh)}$

D. independent of height and gravitational field strength

MCQ 38: Dimensions of Young's modulus are

A. $[M]^{-1} [L]^{-1} [T]^{-2}$

B. $[M]^{-1} [L]^{-2} [T]^{-2}$

C. $[M] [L]^{-2} [T]^{-2}$

D. $[M] [L]^{-1} [T]^{-2}$

MCQ 39: Kepler's 3rd law states that...

A. the periods of revolution of the planets are proportional to the cube of their mean distances from sun

B. the periods of revolution of the planets are inversely proportional to the cube of their mean distances from sun

C. the squares of the periods of revolution of the planets are proportional to the cube of their mean distance from sun

D. the squares of the periods of revolution of the planets are inversely proportional to the cube of their mean distance from sun

Chapter 20

Medical Imaging MCQs

MCQ 1: Gradual decrease in x-ray beam intensity as it passes through material is called

A. attenuation

B. decay

C. radioactivity

D. imaging

MCQ 2: Attenuation coefficient of bone is 600 m^{-1} for x-rays of energy 20 keV and intensity of beam of x-rays is 20 Wm^{-2}, then intensity of beam after passing through a bone of 4mm is

A. 3 Wm^{-2}

B. 2.5 Wm^{-2}

C. 2.0 Wm^{-2}

D. 1.8 Wm^{-2}

MCQ 3: For protons, the gyromagnetic ratio has the value

A. 3×10^8 rads^{-1} T^{-1}

B. 2.68×10^8 rads^{-1} T^{-1}

C. 4×10^8 rads^{-1} T^{-1}

D. 5×10^8 rads^{-1} T^{-1}

MCQ 4: Energy passing through unit area is

A. intensity of x-ray

B. frequency of x-ray

C. wavelength of x-ray

D. amplitude of x-ray

MCQ 5: speed of ultrasound depends upon

A. medium

B. amplitude

C. material

D. wavelength

MCQ 6: Bones look white in x-ray photograph because

A. they are bad absorbers of x-rays

B. they reflect x-rays

C. they are good absorbers of x-rays

D. they are bad absorbers of ultraviolet rays

MCQ 7: Larmor frequency depends upon the

A. individual nucleus

B. magnetic flux density

C. both A and B

D. energetic flux unit

MCQ 8: Acoustic impedance of human skin is

A. 1.65×10^6 kg m² s⁻¹

A. 1.65×10^6 kg m^2 s^{-1}

B. 1.71×10^6 kg m^{-2} s^{-1}

C. 2×10^6 kg m^{-2} s^{-1}

D. 2×10^7 kg m^{-3} s^{-2}

MCQ 9: In the best piezo-electric substances, the maximum value of strain is about

A. 0.50%

B. 0.40%

C. 0.30%

D. 0.10%

MCQ 10: With gel between skin and transducer percentage of reflected intensity of ultrasonic is

A. 0.03%

B. 0.05%

C. 0.06%

D. 0.08%

MCQ 11: Attenuation coefficient depends on

A. frequency of x-ray photons

B. wavelength of x-ray photons

C. energy of x-ray photons

D. amplitude of x-ray photons

MCQ 12: x-rays are filtered out of human body by using

A. cadmium absorbers

B. carbon absorbers

C. copper absorbers

D. aluminum absorbers

MCQ 13: Wavelength of x-rays is in range

A. 10^{-8} to 10^{-13} m

B. 10^{-7} to 10^{-14} m

C. 10^{-10} to 10^{-15} m

D. 10^2 to 10^9 m

MCQ 14: If fast moving electrons rapidly decelerate, then rays produced are

A. alpha rays

B. beta rays

C. gamma rays

D. x-rays

MCQ 15: As the x-rays pass through matter, it's intensity

A. increases

B. decreases

C. remains constant

D. may increase or decrease depending on the object

MCQ 16: X-rays have

A. short wavelength

B. high frequency

C. both A and B

D. longest wavelength

MCQ 17: Acoustic impedance is defined as

A. ϱ/c

B. ϱc

C. c/ϱ

D. $\varrho+c$

MCQ 18: A sound wave which has frequency higher than the upper limit of human hearing is

A. infra sonic

B. ultrasonic

C. supersonic

D. megasonic

MCQ 19: Fatty tissues have

A. relaxation time of several seconds

B. relaxation time of several hundred nanoseconds

C. intermediate relaxation times

D. relaxation times of several hundred milliseconds

MCQ 20: Angular frequency of precision is called

A. Lower frequency

B. higher frequency

C. Larmor frequency

D. linear frequency

MCQ 21: Scattered x-ray beams approach the detector screen

A. perpendicularly

B. parallel

C. anti-parallel

D. at an angle

MCQ 22: Maximum energy an x-ray photon can have is

A. e/V

B. e

C. eV

D. V

MCQ 23: Soft x-rays have

A. high energies

B. low energies

C. lowest frequency

D. longest wavelength

MCQ 24: Intensity of x-rays can be increased by increasing

A. frequency

B. current

C. voltage

D. resistance

MCQ 25: Bone thickness is equal to

A. $c\Delta t/2$

B. $c\Delta t$

C. c/t

D. $t/2$

MCQ 26: Hardness of x-ray beam can be increased by increasing

A. voltage

B. current

C. frequency

D. wavelength

MCQ 27: In x-ray production, the kinetic energy of an electron arriving at anode is

A. 100 keV

B. 200 keV

C. 300 keV

D. 400 keV

MCQ 28: Change in speed of ultrasound causes

A. reflection

B. diffraction

C. refraction

D. image

MCQ 29: Type of x-rays used to detect break in bone is

A. hard

B. soft

C. both A and B

D. moderate

MCQ 30: Wavelength of 2.0 MHz ultrasound waves in tissue is

A. 7.5×10^{-4} m

B. 8×10^{-5} m

C. 8.5×10^{-6} m

D. 9.2×10^{-3} m

MCQ 31: Intensifier screens reduces the patient's exposure to x-rays by a factor of

A. 500-600

B. 1000-2000

C. 100-500

D. 10-100

MCQ 32: Contrast media consist of elements with

A. lower atomic number

B. higher atomic number

C. metalloids

D. inert gases

MCQ 33: Thickness of material which decreases intensity of x-ray material to half of original value is

A. quarter thickness

B. half thickness

C. half life

D. 2 times of thickness

MCQ 34: A good x-ray source should produce x-rays of narrow beam and

A. parallel x-rays

B. perpendicular x-rays

C. anti-parallel x-rays

D. anti-perpendicular x-rays

Chapter 21

Momentum MCQs

MCQ 1: Speed of Earth when a rock of mass 60 kg falling towards Earth with speed of 20 ms^{-1} is

A. 2.4×10^{-22} ms^{-1}

B. 3.5×10^{-33} ms^{-1}

C. -2.0×10^{-22} ms^{-1}

D. -3×10^{34} ms^{-1}

MCQ 2: Force exerted by bat on ball if it strikes a ball of mass 0.16 kg initially hits bat with speed of 25 ms^{-1} with time impact of 0.003 s is

A. 145 N

B. 1333.33 N

C. 1456.7 N

D. 6543 N

MCQ 3: Momentum of electron having mass 9.1×10^{-31} kg and velocity 2.0×10^7 is

A. 1.91×10^{-23}

B. 2.34×10^{-23}

C. 3.11×10^{-19}

D. 7.88×10^{-34}

MCQ 4: To replace a ball with another ball by collision, a snooker player must consider the condition that

A. the collision must be head on

B. The moving ball must not be given any spin

C. both A and B

D. no conditions required

MCQ 5: Direction of momentum is direction of object's

A. mass

B. acceleration

C. velocity

D. frictional force

MCQ 6: In a springy collision, if the fast moving trolley collides with a slow one, then the fast one will bounce back at speed of

A. slow one

B. less than slow one

C. more than slow one

D. with the same speed as before

MCQ 7: An object travelling with constant velocity has

A. constant momentum

B. zero momentum

C. increasing momentum

D. decreasing momentum

MCQ 8: If a trolley collides with a stationary trolley of double mass, then they move off with

A. half of the original velocity

B. one third of original velocity

C. double the original velocity

D. triple the original velocity

MCQ 9: In a perfectly elastic collision, momentum and energy are

A. not conserved

B. conserved

C. becomes zero after collision

D. equal before collision

MCQ 10: Mass and velocity are combined to give

A. angular momentum

B. equilibrium

C. acceleration

D. linear momentum

MCQ 11: If the total kinetic energy and momentum of a system becomes zero after collision, then the collision is

A. elastic

B. inelastic

C. conserved

D. not conserved

MCQ 12: Momentum of two objects moving with same speed but in opposite direction upon collision is

A. increased

B. decreased

C. is zero

D. is infinite

MCQ 13: Forces on interacting bodies are

A. equal

B. opposite

C. both A and B

D. parallel

MCQ 14: Resultant force acting on object and rate of change of linear momentum are

A. inversely related

B. not related at all

C. directly related

D. directly proportional

MCQ 15: In a perfectly elastic collision, the relative speed of approach and relative speed of separation are

A. equal

B. in equal

C. zero

D. infinite

MCQ 16: A white ball of mass 1.0 kg moving with initial speed $u = 0.5$ ms^{-1} collides with stationary red ball of same mass, they move forward making angle of $90°$ between their paths. Their speed is

A. 1 ms^{-1}

B. 0.354 ms^{-1}

C. 2 ms^{-1}

D. 3 ms^{-1}

MCQ 17: In fireworks the momentum provided by chemicals is directed

A. upwards

B. left side

C. right side

D. downwards

MCQ 18: Total momentum within a closed system is

A. increasing

B. decreasing

C. zero

D. constant

MCQ 19: Average force acting on 900 kg car if its velocity ranges from 5 ms^{-1} to 30 ms^{-1} in 12 s is

A. 1875 N

B. 2000 N

C. 3000 N

D. 1560 N

MCQ 20: Interaction that causes an object's momentum to change is

A. velocity

B. acceleration

C. power

D. force

MCQ 21: Momentum possessed by spinning objects is called

A. linear momentum

B. angular momentum

C. normal momentum

D. degrees' momentum

MCQ 22: In a perfectly inelastic collision, kinetic energy

A. totally disappears

B. is increased

C. is decreased

D. is unchanged

Chapter 22

Motion Dynamics MCQs

MCQ 1: Contact force always acts at

A. acute angles to the surface producing it

B. right angles to the surface producing it

C. obtuse angle to the surface producing it

D. parallel to the surface producing it

MCQ 2: Combinations of base units are

A. simple units

B. derived units

C. scalars

D. vectors

MCQ 3: Two forces which make up Newton's third law can

A. act on same objects

B. act on different objects

C. not act at same time

D. not act oppositely

MCQ 4: Rate of falling object in vacuum is

A. independent of weight

B. dependent on mass

C. independent of mass

D. dependent of weight

MCQ 5: At terminal velocity the

A. air resistance and weight are equal

B. air resistance is less than weight

C. weight is more than air resistance

D. air resistance is more than weight

MCQ 6: Vehicle will accelerate as long as

A. air resistance is greater than thrust

B. air resistance is greater than inertia

C. thrust is greater than air resistance and friction

D. friction is greater than thrust

MCQ 7: 1 N is equal to

A. 1 kg ms^{-2}

B. 10 kg ms^{-1}

C. 10 kg ms^{-2}

D. 100 kg ms^{-2}

MCQ 8: Density of air is

A. 1/8 of water

B. 1/7 of water

C. 1/45 of water

D. 1/800 of water

MCQ 9: Until a force acts on a body, its velocity is

A. zero

B. constant

C. increasing

D. decreasing

MCQ 10: If there is no net force acting on body, then its acceleration is

A. zero

B. constant

C. increasing

D. decreasing

MCQ 11: Acceleration of a rocket having mass 5000 kg and resultant force acting on it is 200,000 N is

A. 50 ms^{-2}

B. 56 ms^{-2}

C. 70 ms^{-2}

D. 40 ms^{-2}

MCQ 12: Force which makes it difficult to run through shallow water is called

A. viscosity

B. up thrust

C. friction

D. drag

MCQ 13: Prefix for 10^{-9} is

A. micro

B. deci

C. centi

D. nano

MCQ 14: Acceleration due to gravity on moon is

A. 9.9 ms^{-2}

B. 9.5 ms^{-2}

C. 6.1 ms^{-2}

D. 1.6 ms^{-2}

MCQ 15: An object immersed in fluid experiences an upward force named

A. viscosity

B. drag force

C. up thrust

D. friction

MCQ 16: If each term in an equation has same base units then the equation is said to be

A. homogeneous

B. non homogeneous

C. equation of straight line

D. equation of circle

MCQ 17: A force similar to friction is

A. forward force

B. pulling force

C. drag force

D. contact force

MCQ 18: Forces acting on an object are balanced if resultant force on object is

A. constant

B. zero

C. increasing

D. decreasing

MCQ 19: Base unit among following is

A. Newton

B. Joule

C. Candela

D. Watt

MCQ 20: Force provided by breaking system of train if it is decelerating at rate of -3 ms^{-2} and having mass 10,000 kg is

A. -30,000 N

B. -40,000 N

C. -50,000 N

D. 30,000 N

MCQ 21: Force applied on a body and its acceleration are

A. inversely related

B. directly related

C. not related at all

D. inversely proportional

MCQ 22: The larger the mass of a moving object the

A. larger the acceleration produced

B. acceleration becomes constant

C. smaller the acceleration

D. acceleration becomes zero

MCQ 23: 500 MW can be written in powers of 10 as

A. 500×10^6

B. 500×10^3

C. 500×10^{-6}

D. 500×10^9

MCQ 24: Point where entire weight of an object acts is

A. edge

B. center of gravity

C. central point

D. can be anywhere in body

MCQ 25: Another name for force of gravity acting on an object is

A. friction

B. air resistance

C. weight

D. mass

MCQ 26: When two objects are in contact, they exert forces in

A. opposite direction

B. same directions

C. can be both A and B

D. perpendicular direction

Chapter 23

Nuclear Physics MCQs

MCQ 1: Activity is proportional to number of

A. daughter nuclei

B. decayed nuclei

C. undecayed nuclei

D. father nuclei

MCQ 2: Energy given to nucleus to dismantle it increases the

A. kinetic energy of individual nucleons

B. mechanical energy of individual nucleons

C. potential energy of individual nucleons

D. chemical energy of individual nucleons

MCQ 3: Radioactive decay is a

A. random process

B. non-spontaneous process

C. regular process

D. massive process

MCQ 4: 1 u is equal to

A. 1.660×10^{-27} kg

B. 2×10^{-27} kg

C. 3×10^{-27} kg

D. 5×10^{-27} kg

MCQ 5: In gamma emission, the change in nucleon number is

A. zero

B. definite

C. increase by 1

D. decreases by 1

MCQ 6: At higher energy, the bodies have

A. small mass

B. large mass

C. zero mass

D. smaller weight

MCQ 7: Time taken by a radioactive substance to decay half is called

A. time delay

B. half life

C. time constant

D. half period

MCQ 8: Most stable isotope in nature is of

A. iron-56

B. carbon-12

C. uranium-235

D. uranium-238

MCQ 9: Activity of one decay per second is equal to

A. 1 Bq

B. 1 atm

C. 1 mol

D. 1 Cd

MCQ 10: The greater the decay constant

A. the less the activity

B. the greater the activity

C. the greater the size

D. the less the size

MCQ 11: Total amount of mass and energy together in a system is

A. increasing

B. decreasing

C. zero

D. constant

MCQ 12: Process by which energy is released in sun is

A. fission

B. Haber's process

C. fusion

D. radioactivity

MCQ 13: Minimum energy required to pull nucleus apart is called

A. ionization energy

B. electron affinity

C. chemical energy

D. binding energy

MCQ 14: Mass excess for U-235 is

A. 0.034 u

B. 0.043 u

C. 0.05 u

D. 0.06 u

MCQ 15: As compared to proton, mass of neutron is

A. 10% greater

B. 5% greater

C. 1% greater

D. 0.1% greater

MCQ 16: 1 mole of uranium-238 has potential to emit total energy equal to about

A. 10^9 J

B. 10^{10} J

C. 10^{11} J

D. 10^{12} J

MCQ 17: If energy is released from a system, it's mass

A. decreases

B. increases

C. constant

D. zero

MCQ 18: New nucleus after alpha particle decays, is called

A. parent nucleus

B. daughter nucleus

C. decayed nucleus

D. undecayed nucleus

MCQ 19: If nucleus is formed from separate nucleons, then energy is

A. gained

B. released

C. converted

D. absorbed

Chapter 24

Oscillations MCQs

MCQ 1: Maximum displacement from equilibrium position is

A. frequency

B. amplitude

C. wavelength

D. period

MCQ 2: Displacement-time graph depicting an oscillatory motion is

A. cos curve

B. sine curve

C. tangent curve

D. straight line

MCQ 3: In SHM, velocity at equilibrium position is

A. minimum

B. constant

C. maximum

D. zero

MCQ 4: Natural frequency of a guitar string can be changed by changing it's

A. area

B. diameter

C. length

D. stiffness

MCQ 5: Over-damping results in

A. slower return to equilibrium

B. faster return to equilibrium

C. equilibrium is never achieved

D. arrhythmic return to equilibrium

MCQ 6: Our eyes detect the oscillations up to

A. 8 Hz

B. 9 Hz

C. 6 Hz

D. 5 Hz

MCQ 7: For SHM, maximum speed is proportional to

A. wavelength

B. acceleration

C. time

D. frequency

MCQ 8: A force that acts to return the mass to its equilibrium position is called

A. frictional force

B. restoring force

C. normal force

D. contact force

MCQ 9: In cars, springs are damped by

A. shock absorbers

B. engine

C. tyres

D. brake pedals

MCQ 10: If time period of an oscillation is 0.40 s, then its frequency is

A. 2 Hz

B. 2.5 Hz

C. 3 Hz

D. 3.5 Hz

MCQ 11: As amplitude of resonant vibrations decreases, degree of damping

A. increases

B. remains same

C. decreases

D. varies

MCQ 12: Oscillations become damped due to

A. normal force

B. friction

C. tangential force

D. parallel force

MCQ 13: In SHM, object's acceleration depends upon

A. displacement from equilibrium position

B. magnitude of restoring force

C. both A and B

D. force exerted on it

MCQ 14: angular frequency of SHM is equal to

A. 2π

B. $2\pi f$

C. $2f$

D. $1/T$

MCQ 15: For a resonating system it should oscillate

A. bound

B. only for some time

C. freely

D. for infinite time

MCQ 16: Velocity at equilibrium position is

A. constant

B. minimum

C. maximum

D. zero

MCQ 17: If the swing moves from right to left, then velocity is

A. negative

B. positive

C. constant

D. zero

MCQ 18: Acceleration is directly related to

A. displacement

B. negative of displacement

C. velocity

D. negative of speed

MCQ 19: Gradient of velocity-time graph gives

A. force

B. frequency

C. wavelength

D. acceleration

MCQ 20: Magnitude of gradient of a-x graph is

A. ω

B. ω^2

C. $\omega/2$

D. ω^3

MCQ 21: Potential energy of system is maximum at

A. extreme position

B. mean position

C. in between extreme and mean position

D. moderate position

MCQ 22: In SHM, acceleration is always directed towards the

A. equilibrium position

B. mean position

C. tangent to the motion

D. downwards

MCQ 23: Number of oscillations per unit time is

A. amplitude

B. wavelength

C. frequency

D. period

MCQ 24: When displacement x = 0, then kinetic energy of system is

A. minimum

B. maximum

C. constant

D. zero

MCQ 25: Energy of a system executing SHM is

A. increasing

B. decreasing

C. constant

D. variable

MCQ 26: Oscillatory motion has a

A. straight lined graph

B. randomly lined graph

C. sinusoidal graph

D. asymptotic graph

MCQ 27: During one oscillation, phase of oscillation changes by

A. π rad

B. $\pi/2$ rad

C. 2π rad

D. 4π rad

MCQ 28: If an object moves back and forth repeatedly around a mean position it is called

A. oscillating

B. revolving

C. rotating

D. motion

Chapter 25

Physics Problems MCQs

MCQ 1: Terminal potential difference of battery depends on

A. current

B. temperature

C. both A and B

D. resistance of external resistor

MCQ 2: Systematic errors occur due to

A. overuse of instruments

B. careless usage of instruments

C. both A and B

D. human sight

MCQ 3: In cells, internal resistance is due to

A. components inside

B. chemicals within

C. lead blocks

D. graphite

MCQ 4: Measurement which is close to the true value is

A. accurate

B. average

C. precise

D. error

MCQ 5: A train with a whistle that emits a note of frequency 800 Hz is approaching an observer at speed of 60 ms^{-1}, the frequency of note heard by observer is

A. 978 Hz

B. 980 Hz

C. 950 Hz

D. 900 Hz

MCQ 6: Smallest division on a rule is of

A. 1 cm

B. 1 m

C. 1 mm

D. 10 cm

MCQ 7: If the two quantizes are inversely proportional then graph between y and 1/x gives

A. straight line cutting the axis

B. straight line through origin

C. a parabolic curve

D. a hyperbolic curve

MCQ 8: Systematic errors can be removed by

A. buying new instrument

B. breaking the instrument

C. dusting the instrument

D. recalibrating the instrument

MCQ 9: Least count of screw gauge is

A. 0.01 cm

B. 0.5 cm

C. 0.1 cm

D. 0.01 mm

MCQ 10: Increase in kinetic energy of car having mass 800 kg and velocities ranging from 20 ms^{-1} to 30 ms^{-1} is

A. 300 KJ

B. 500 KJ

C. 400 KJ

D. 200 KJ

MCQ 11: Work done by a person having weight 600 N and he needs to climb up a mountain of height 2000 m is

A. 1000 kJ

B. 1300 kJ

C. 1400 kJ

D. 1200 kJ

MCQ 12: Internal resistance of cell when there is current of 0.40 A when a battery of 6.0 V is connected to a resistor of 13.5 Ω is

A. 1.5 Ω

B. 2.3 Ω

C. 3.5 Ω

D. 4.5 Ω

MCQ 13: Barrel of screw gauge has

A. 100 divisions

B. 50 divisions

C. 10 divisions

D. 45 divisions

MCQ 14: A measurement which on repetition gives same or nearly same result is called

A. accurate measurement

B. average measurement

C. precise measurement

D. estimated measurement

MCQ 15: A motor of the lift provides a force of 20 kN which rises it by 18 m in 10 s, the output power of motor is

A. 36 kW

B. 46 kW

C. 56 kW

D. 66 kW

MCQ 16: Actual range of values around a measurement is called

A. error

B. uncertainty

C. accuracy

D. precision

MCQ 17: Depth of water in a bottle is 24.3 cm and uncertainty is 0.2 cm, percentage uncertainty in measurement is

A. 0.82%

B. 9%

C. 1%

D. 2%

MCQ 18: While taking the reading, the line of sight should be

A. at acute angle to the scale

B. perpendicular to scale

C. at obtuse angle to the scale

D. parallel to the scale

MCQ 19: Maximum current a battery of EMF 3.0 V and internal resistance 1.0 Ω is

A. 4.0 A

B. 5.0 A

C. 3.0 A

D. 30 A

MCQ 20: In potential dividers, the output voltage depends upon the

A. single resistor

B. relative values of all resistors

C. current

D. temperature

MCQ 21: Calipers are used to measure the

A. diameter

B. length

C. thickness

D. volume

MCQ 22: Smallest division on stopwatch is

A. 0.1 s

B. 0.05 s

C. 0.01 s

D. 1 s

Chapter 26

Physics: Waves MCQs

MCQ 1: intensity of sun's radiation is about

A. 1.0 kW m^{-2}

B. 20 kW m^{-2}

C. 5 kW m^{-2}

D. 8 kW m^{-2}

MCQ 2: When light enters from vacuum in to glass, its velocity

A. decreases

B. remains same

C. increases

D. varies depending on mass of glass

MCQ 3: As wave travels, intensity

A. increases

B. remains same

C. decreases

D. varies

MCQ 4: Waves that move through materials are called

A. progressive waves

B. EM waves

C. radio waves

D. UV waves

MCQ 5: Speed of sound in air is

A. 280 ms^{-1}

B. 300 ms^{-1}

C. 350 ms^{-1}

D. 330 ms^{-1}

MCQ 6: Particles vibrate about their mean positions and transfer

A. frequency

B. wavelength

C. energy

D. power

MCQ 7: Speed of electromagnetic radiation is independent of

A. wavelength

B. amplitude

C. time period

D. frequency

MCQ 8: Mechanical waves include

A. sound

B. light

C. EM waves

D. UV waves

MCQ 9: A complete cycle of wave is around

A. 90°

B. 180°

C. 45°

D. 360°

MCQ 10: Changing magnetic field induces

A. charge

B. current

C. frequency

D. voltage

MCQ 11: Phase difference is measured in

A. degrees

B. meters

C. seconds

D. newton

MCQ 12: Speed at which stars and galaxies are moving away from us is determined by phenomena of

A. blue shift

B. yellow shift

C. red shift

D. orange shift

MCQ 13: Electric and magnetic fields vary at angle of

A. 30°

B. 90°

C. 45°

D. 180°

MCQ 14: Unification of electromagnetic and weak nuclear forces was done by

A. Maxwell

B. Faraday

C. Kirchhoff

D. Abdus-Salam

MCQ 15: Frequency and time period are

A. directly related

B. not related

C. inversely related

D. directly proportional

MCQ 16: Amount by which one oscillation leads or lags behind another is called

A. in phase

B. intensity

C. phase difference

D. superposition

MCQ 17: Number of oscillations per unit time is called

A. wavelength

B. amplitude

C. displacement

D. frequency

MCQ 18: If the particles of medium vibrate at right angles to the direction of velocity then wave is

A. longitudinal

B. transverse

C. abrupt

D. sound

MCQ 19: Intensity of a wave is directly proportional to the

A. amplitude

B. square of amplitude

C. cube of amplitude

D. frequency

MCQ 20: Wave speed is directly proportional to

A. frequency

B. amplitude

C. wavelength

D. energy

MCQ 21: Longitudinal waves gives rise to

A. amplitude

B. frequency

C. wavelength

D. high and low pressure regions

MCQ 22: Ranges of waves which overlap are

A. x-rays and gamma rays

B. x-rays and infrared rays

C. gamma rays and infrared rays

D. UV rays and infrared rays

Chapter 27

Quantum Physics MCQs

MCQ 1: In order to find the internal structure of nucleus, electrons should be accelerated by voltages up to

A. 10^5 V

B. 10^7 V

C. 10^9 V

D. 10^{11} V

MCQ 2: High speed electrons have wavelength of order

A. 10^{-15} m

B. 10^{-14} m

C. 10^{-16} m

D. 10^{-17} m

MCQ 3: Wavelength of slow moving neutrons is about

A. 10^{-34} m

B. 10^{-20} m

C. 10^{-19} m

D. 10^{-10} m

MCQ 4: High speed electrons from particle detectors are used to determine

A. arrangement of atoms in metals

B. diameter of atomic nuclei

C. inter atomic distance

D. circumference of atomic nuclei

MCQ 5: Energy of gamma photon is greater than

A. 10^2 J

B. 10^{-13} J

C. 10^{13} J

D. 10^5 J

MCQ 6: Waves associated with electrons are referred to as

A. plasma waves

B. UV waves

C. gamma rays

D. matter waves

MCQ 7: Frequency below which no electrons are emitted from metal surface is

A. minimum frequency

B. angular frequency

C. maximum frequency

D. threshold frequency

MCQ 8: Loss of energy of an electron results in

A. absorption of photon

B. emission of photon

C. destruction of photon

D. formation of photon

MCQ 9: Wavelength of a 65 kg person running at a speed of 3 ms^{-1} through an opening of width 0.80 m is

A. 34 m

B. 35 m

C. 3.4×10^{-36} m

D. 3.5 m

MCQ 10: According to Newton, the light travels as

A. particles

B. waves

C. both A and B

D. dust

MCQ 11: In electron diffraction, the rings behave as

A. particles

B. waves

C. both A and B

D. rays

MCQ 12: Energy absorbed by electron is used in

A. escaping the metal

B. increasing kinetic energy

C. both A and B

D. increasing frequency

MCQ 13: Diffraction of slow moving electrons is used to estimate

A. arrangement of atoms in metals

B. nature of atoms

C. number of atoms in metals

D. position of atoms in metalloids

MCQ 14: Energy of photon is directly related to the

A. wavelength

B. wave number

C. frequency

D. amplitude

MCQ 15: When a charged particle is accelerated through a potential difference V, its kinetic energy

A. decreases

B. remains same

C. increases

D. varies depending on resistance of wire

MCQ 16: Energy of an electron in an atom is

A. quantized

B. continuous

C. radial

D. randomized

MCQ 17: In dark, LDR has

A. low resistance

B. high current

C. high resistance

D. both A and B

MCQ 18: 1 eV is equal to

A. 1.6×10^{-19} J

B. 2.0×10^{-20} J

C. 3 J

D. 4 J

MCQ 19: Electrons show diffraction effects because their de Broglie wavelength is similar to

A. spacing between atomic layers

B. no. of atomic layers

C. nature of atomic layers

D. positioning of atomic layers

MCQ 20: Characteristic properties of waves are

A. reflection

B. refraction

C. interference

D. all of above

MCQ 21: Plank's constant has units

A. J

B. s

C. J s^{-1}

D. J s

MCQ 22: Gas atoms that exert negligible electrical forces on each other are

A. molecules

B. compounds

C. isotopes

D. isolated atoms

MCQ 23: Wavelength of red color is about

A. 7×10^{-7} m

B. 7×10^{7} nm

C. 4×10^{-7} m

D. 4×10^{-7} nm

MCQ 24: Quantum of electromagnetic energy is called

A. particles

B. photons

C. waves

D. energy

MCQ 25: In photoelectric effect, electrons should be removed from the

A. inner shells

B. surface

C. from core

D. the nucleus

MCQ 26: Light interacts with matter as

A. wave

B. particle

C. both A and B

D. rays

MCQ 27: When white light is passed through cool gases, the spectra observed is called

A. line spectra

B. continuous spectra

C. emission line spectra

D. absorption line spectra

MCQ 28: Wavelength of ultraviolet region of electromagnetic spectrum is

A. 121 nm

B. 120 nm

C. 119 nm

D. 130 nm

MCQ 29: In an insulator, the valence band is

A. fully occupied

B. fully empty

C. half filled

D. half charged

MCQ 30: Most energetic photons are

A. alpha

B. beta

C. gamma

D. x-rays

Chapter 28

Radioactivity MCQs

MCQ 1: Radius of nucleus ranges from

A. 10^{-15} m

B. 10^{-15} m to 10^{-14} m

C. 10^{-10} m

D. 10^{-10} m to 10^{-6} m

MCQ 2: Number of protons in an atom determine

A. chemical properties

B. physical properties

C. magnetic properties

D. electrical properties

MCQ 3: In β^+ decay, an UP quark becomes

A. a strange quark

B. a simple quark

C. a down quark

D. an anti-quark

MCQ 4: Most of the space in an atom is

A. filled with positive charge

B. empty

C. filled with negative charge

D. filled with neutrons

MCQ 5: A proton is made up of

A. one up quark and two down quarks

B. an up quark and down antiquark

C. two up quarks and a down quark

D. strange quark and an anti-strange quark

MCQ 6: Neutrinos have electric charge of

A. zero

B. 1

C. 2

D. 3

MCQ 7: β^- radiations are simply

A. protons

B. neutrons

C. electrons

D. muons

MCQ 8: In a nuclear process, the quantity conserved is

A. mass-energy

B. momentum

C. mass only

D. energy only

MCQ 9: A specific combination of protons and neutrons in a nucleus is called

A. nucleons

B. nuclide

C. neutrino

D. nucleolus

MCQ 10: In β^+ decay, the nucleon number is

A. conserved

B. not conserved

C. unstable

D. stable

MCQ 11: Particles that are unaffected by strong nuclear force are

A. protons

B. leptons

C. neutrons

D. bosons

MCQ 12: Number of isotopes neon has is

A. 2

B. 4

C. 3

D. 5

MCQ 13: Phenomena of radioactivity was discovered by Henri Becquerel in

A. 1896

B. 1895

C. 1894

D. 1893

MCQ 14: Radiations emitted by radioactive substances is

A. alpha

B. beta

C. gamma

D. all of above

MCQ 15: Elements undergo radioactive decay when proton number becomes greater than

A. 50

B. 40

C. 83

D. 73

MCQ 16: Process in which α and β rays pass close to atoms and knock electrons out is called

A. atomization

B. ionization

C. decay

D. hydroxylation

MCQ 17: Alpha particles have relatively

A. low kinetic energies

B. high potential energy

C. high mechanical energy

D. high kinetic energy

MCQ 18: Strongest ionizing radiation is

A. Alpha

B. beta

C. gamma

D. x-rays

MCQ 19: The nucleon number consists of

A. Number of electrons

B. Number of protons

C. Number of electrons and protons

D. Number of protons and neutrons

MCQ 20: Gamma radiation is emitted in order to

A. excite the atom

B. release excess energy from atom

C. destabilize the atom

D. stabilize the atom

MCQ 21: Electrons move around nucleus in form of

A. clouds

B. dust particles

C. steam

D. charge

MCQ 22: Particles like kaons and muons etc., were found out by

A. looking at cosmic rays

B. looking at particles in accelerators

C. looking closely at atom

D. both A and B

MCQ 23: Type of rays that affect the nucleus are

A. alpha

B. beta

C. gamma

D. EM

MCQ 24: Force that acts on both quarks and leptons is

A. strong nuclear force

B. weak interaction

C. intermediate interaction

D. nuclear force

MCQ 25: Particles that experience strong force are

A. leptons

B. hadrons

C. both A and B

D. softons

MCQ 26: Heavy nuclei have

A. more protons than neutrons

B. more electrons than neutrons

C. more neutrons than electrons

D. more neutrons than protons

MCQ 27: The strong nuclear force acts over the distance

A. 10^{-13} m

B. 10^{-14} m

C. 10^{-15} m

D. 10^{-16} m

MCQ 28: A decay in which a proton decays in to neutron and an electron neutrino is

A. β^+ decay

B. β^-

C. γ decay

D. α decay

MCQ 29: Photon of electromagnetic radiation is

A. α ray

B. β ray

C. γ ray

D. x-ray

MCQ 30: Mass of alpha particle is

A. 50 times the mass of electron

B. 100 times the mass of electron

C. 500 times the mass of electron

D. 1000 times the mass of electron

MCQ 31: Particle which explains about mass of matter is called

A. Higgs boson

B. protons

C. leptons

D. neutrons

MCQ 32: Plum pudding model describes atom as

A. negative pudding with positive plums

B. negative pudding

C. positive pudding with negative plums

D. positive pudding only

MCQ 33: Lepton among them are

A. electrons

B. neutrinos

C. protons

D. both A and B

MCQ 34: Density of proton is equal to density of

A. electron

B. atom

C. neutron

D. neutrino

Chapter 29

Resistance and Resistivity MCQs

MCQ 1: Resistivity of lead is

A. 22.5×10^{-8} Ω m

B. 20.8×10^{-8} Ω m

C. 10 Ω m

D. 5 Ω m

MCQ 2: A filament lamp is

A. Ohmic

B. non-Ohmic

C. low resistive

D. non glowing

MCQ 3: In case of filament lamp at higher voltages, the resistance of lamp

A. decreases

B. increases

C. remains constant

D. varies depending on the filament

MCQ 4: If the connections across the resistor are reversed, then graph between current and potential difference is

A. variable

B. straight lined

C. constant

D. inverted

MCQ 5: In semiconductors upon increasing temperature, conductivity

A. decreases

B. increases

C. remains constant

D. haphazard

MCQ 6: In NTC thermistor on increasing temperature, the resistance

A. increases

B. remains constant

C. decreases

D. behaves abruptly

MCQ 7: A component that allows unidirectional current to pass through it is

A. resistor

B. inductor

C. transformer

D. diode

MCQ 8: A rectifier converts

A. DC to AC

B. AC to DC

C. voltage in to current

D. current in to voltage

MCQ 9: Threshold voltage at which LED emits light is

A. less than 0.6

B. equal to 0.6

C. more than 0.6

D. more than 3

MCQ 10: Resistivity is measured in

A. ohms

B. ohm per meter

C. ohm meter

D. ohm square meter

MCQ 11: Component which obeys Ohm's law is called

A. resistive component

B. efficient component

C. Ohmic component

D. non-Ohmic component

MCQ 12: At constant temperature, resistance and cross-sectional area are

A. directly related

B. not related

C. remains constant

D. inversely related

MCQ 13: Current and voltage are

A. directly related

B. inversely related

C. not related

D. gives abnormal behavior on graph

MCQ 14: At threshold voltage, resistance of diode

A. decreases

B. increases

C. remains constant

D. varies depending on the direction of voltage

MCQ 15: Resistance of a particular wire depends upon

A. size and shape

B. shape and length only

C. size only

D. shape only

MCQ 16: Resistance of metal is affected by

A. presence of impurities

B. temperature

C. both A and B

D. pressure

MCQ 17: If current and potential difference are directly related then object follows

A. Ohm's law

B. Faraday's law

C. Ampere's law

D. Kirchhoff's law

Chapter 30

Superposition of Waves MCQs

MCQ 1: Effect of diffraction is greatest if waves pass through a gap with width equal to

A. frequency

B. wavelength

C. amplitude

D. wavefront

MCQ 2: Visible light has wavelength of

A. 5×10^{-7} m

B. 3×10^{8} m

C. 6×10^{3} m

D. 4×10^{4} m

MCQ 3: From double-slit experiment, the quantities to be measured are

A. slit separation

B. fringe separation

C. slit-to-screen distance

D. all of above

MCQ 4: For destructive interference, the path difference is

A. odd number of half wavelengths

B. even number of half wavelengths

C. whole number of wavelengths

D. even whole number of wavelengths

MCQ 5: Constructive interference happens when two waves are

A. out of phase

B. zero amplitude

C. in phase

D. in front

MCQ 6: Two waves with phase difference 180° have resultant of amplitude

A. one

B. zero

C. same as the single wave

D. doubles the single wave

MCQ 7: If two waves are in phase and have same amplitude then resultant wave has

A. half of amplitude of single wave

B. same amplitude as single wave

C. twice of amplitude of single wave

D. thrice of amplitude of single wave

MCQ 8: For listening radio in cars, external radio aerials are used because

A. radio waves have shorter wavelength

B. radio waves have longer wavelength

C. radio waves cannot pass through window

D. radio waves require a medium to propagate

MCQ 9: When two waves meet, their displacements

A. add up

B. cancel out

C. destruct each other

D. subtract down

MCQ 10: Splitting of white light in to constituent colors is called

A. diffraction

B. refraction

C. reflection

D. dispersion

MCQ 11: Grating element is equal to

A. $n\lambda/\sin\theta$

B. $n\lambda$

C. $\sin\theta$

D. cosθ

MCQ 12: With diffraction grating, the angles are

A. small

B. greater

C. zero

D. close to zero

MCQ 13: Extra distance travelled by one of waves compared with other is called

A. path

B. displacement

C. phase difference

D. path difference

MCQ 14: Spreading of wave as it passes through a gap or around an edge is called

A. reflection

B. refraction

C. diffraction

D. superposition

MCQ 15: Fringes are referred to as

A. minima

B. maxima

C. nodes

D. normal points

MCQ 16: Principle of superposition can be applied to

A. EM waves

B. sound waves

C. radio waves

D. all of above

MCQ 17: Coherent sources emit waves that have

A. increasing phase difference

B. decreasing phase difference

C. constant phase difference

D. varying phase difference

MCQ 18: Interference pattern of light and dark bands on screen is called

A. graphical pattern

B. line spectrum

C. light spectrum

D. fringes

MCQ 19: Microwaves have wavelength of about

A. 10 cm

B. 20 cm

C. 30 cm

D. 40 cm

MCQ 20: In young double slit experiment, the wavelength of incident beams should be

A. same

B. different

C. zero

D. opposite

MCQ 21: Wavelength of an incident light when it is incident normally on a diffraction grating having 3000 lines per centimeter angular separation is $10°$ is

A. 500 nm

B. 650 nm

C. 580 nm

D. 600 nm

Chapter 31

Thermal Physics MCQs

MCQ 1: Supply of energy depends upon

A. mass of material

B. the change in temperature

C. the material itself

D. all of above

MCQ 2: All substances have minimum internal energy at

A. absolute zero

B. 0°C

C. 0°F

D. 100K

MCQ 3: Specific heat of aluminum when 26400 J of energy is supplied to 2 kg block and it's temperature rises from 20 °C to 35 °C is

A. 1000 J kg^{-1} K^{-1}

B. 70 J kg^{-1} K^{-1}

C. 400 J kg^{-1} K^{-1}

D. 880 J kg^{-1} K^{-1}

MCQ 4: On compression, the gat gets hotter due to

A. increase in kinetic energy

B. decrease in kinetic energy

C. increase in potential energy

D. increase in atomic collisions

MCQ 5: If there is no transfer of energy between two objects then their temperature is

A. same

B. different

C. zero

D. infinite

MCQ 6: Average kinetic energy of gas molecules is proportional to

A. internal energy

B. thermodynamic temperature

C. enthalpy

D. condensation point

MCQ 7: Celsius scale is based on properties of

A. Mercury

B. Aluminum

C. Cesium

D. Water

MCQ 8: Energy required per unit mass of substance to raise temperature of that substance by 1 K is called

A. enthalpy

B. internal energy

C. specific heat capacity

D. temperature

MCQ 9: Measure of average kinetic energy of molecules is

A. temperature

B. energy

C. internal energy

D. enthalpy

MCQ 10: Energy of molecules of any substance is known as

A. kinetic energy

B. potential energy

C. internal energy

D. chemical energy

MCQ 11: Due to evaporation from certain surface, it's temperature

A. falls

B. increases

C. doesn't change

D. becomes zero

MCQ 12: Change of liquid in to gas without boiling is called

A. vaporization

B. sublimation

C. boiling

D. evaporation

MCQ 13: On heating a solid, as the separation of atoms increases the potential energy

A. decreases

B. remains constant

C. increases

D. becomes zero

MCQ 14: Temperature can't be lower than

A. $0\,°C$

B. $0\,°F$

C. $0\,K$

D. $0\,°R$

MCQ 15: When a substance is melted, its temperature doesn't rise because

A. energy is lost somewhere

B. energy is used to break the bonds

C. energy is used to make bonds

D. energy is absorbed to make bonds

Chapter 32

Work, Energy and Power MCQs

MCQ 1: As the object gains speed, it's G.P.E (Gravitational Potential Energy)

A. increases

B. remains constant

C. decreases

D. varies depending on altitude

MCQ 2: If energy loss is zero then decrease in G.P.E is equal to

A. decreases in kinetic energy

B. gain in kinetic energy

C. constant kinetic energy

D. zero kinetic energy

MCQ 3: Average power of all activities of our body is

A. 111 W

B. 113 W

C. 116 W

D. 120 W

MCQ 4: Energy object possesses due to its position is called

A. kinetic energy

B. mechanical energy

C. potential energy

D. chemical energy

MCQ 5: Change in G.P.E can be written as

A. mg h

B. m h

C. mg

D. g h

MCQ 6: Efficiency of car engine is only about

A. 10%

B. 20%

C. 30%

D. 80%

MCQ 7: Work is defined as

A. $F \times S$

B. $F \times a$

C. $m \times a$

D. $F \times m$

MCQ 8: Energy transferred to stone of weight 10 N falling from top of 250 m high cliff is

A. 25000 J

B. 250000 J

C. 2500 J

D. 250 J

MCQ 9: If force of 1 N moves an object through 1 m, then work done is

A. 2 J

B. 1 J

C. 3 J

D. 4 J

MCQ 10: SI units for energy and work are

A. Joules and kg

B. Joules and meter

C. Joules and newton

D. Joules

MCQ 11: If the ball is thrown upwards, the energy changings are

A. K.E changes to G.P.E

B. G.P.E changes to K.E

C. K.E changes to mechanical energy

D. mechanical energy changes to K.E

MCQ 12: 1 W is equal to

A. 10 J s^{-1}

B. 1 J s

C. 1 J s^{-1}

D. 100 J s

MCQ 13: Work done by force of gravity on a satellite of 500 N at height is

A. 0 J

B. 1 J

C. 2 J

D. 3 J

MCQ 14: Rate of doing work is called

A. power

B. energy

C. velocity

D. force

MCQ 15: A weight-lifter raises weights with a mass of 200 kg from the ground to a height of 1.5 m, the increase in their G.P.E is

A. 1962 J

B. 2940 J

C. 800 J

D. 1000 J

Answers Keys

Chapter 1: Accelerated Motion MCQs

1. D	2. D	3. B	4. D	5. B
6. A	7. D	8. B	9. A	10. A
11. B	12. D	13. A	14. D	15. A
16. C	17. D	18. B	19. C	20. C
21. A	22. C			

Chapter 2: Alternating Current MCQs

1. D	2. D	3. C	4. B	5. A
6. D	7. D	8. A	9. C	10. C
11. C	12. B	13. B	14. D	15. A
16. A				

Chapter 3: AS Level Physics MCQs

1. D	2. A	3. B	4. B	5. D
6. B	7. D	8. C	9. D	10. A
11. A	12. A	13. A	14. C	15. B
16. B	17. C	18. D	19. B	20. D
21. B	22. A	23. B	24. A	25. C
26. D	27. C	28. C	29. A	30. A
31. A	32. B	33. A	34. B	35. C

Chapter 4: Capacitance MCQs

1. C	2. A	3. B	4. B	5. A
6. B	7. A	8. C	9. A	10. C
11. C	12. A			

Chapter 5: Charged Particles MCQs

1. A	2. C	3. A	4. D	5. B
6. C	7. A	8. A	9. C	10. C
11. A				

Chapter 6: Circular Motion MCQs

1. C	2. A	3. B	4. C	5. A
6. D	7. D	8. C	9. A	10. B
11. D	12. C	13. A	14. D	15. A
16. B	17. C			

Chapter 7: Communication Systems MCQs

1. B	2. C	3. A	4. A	5. D
6. B	7. D	8. B	9. B	10. C
11. A	12. B	13. A	14. A	15. B
16. A	17. D	18. C	19. C	20. D
21. D	22. B	23. B	24. A	25. D

Chapter 8: Electric Current, Potential Difference and Resistance MCQs

1. D	2. C	3. A	4. A	5. D
6. C	7. C	8. D	9. A	10. B
11. D	12. C	13. A	14. B	15. B
16. B	17. A	18. D	19. D	20. B
21. C	22. B	23. B		

Chapter 9: Electric Field MCQs

1. B	2. A	3. C	4. D	5. A
6. A	7. B	8. A	9. C	10. B
11. D				

Chapter 10: Electromagnetic Induction MCQs

1. D   2. B   3. A   4. A   5. B
6. C   7. C   8. D   9. B   10. D
11. A   12. A   13. C   14. D

Chapter 11: Electromagnetism and Magnetic Field MCQs

1. D   2. C   3. A   4. B   5. D
6. A   7. B   8. B   9. C   10. A
11. D   12. A   13. C   14. D   15. C
16. B   17. B   18. C   19. D

Chapter 12: Electronics MCQs

1. D   2. B   3. D   4. A   5. D
6. C   7. C   8. D   9. A   10. C
11. A   12. C   13. D   14. A   15. B
16. B   17. D   18. B   19. A   20. A
21. A   22. B   23. B   24. B

Chapter 13: Forces, Vectors and Moments MCQs

1. A   2. B   3. B   4. D   5. C
6. A   7. C   8. C   9. C   10. B
11. A   12. B

Chapter 14: Gravitational Field MCQs

1. D   2. A   3. C   4. D   5. A
6. A   7. B   8. B   9. C   10. C
11. B   12. A   13. B   14. A   15. C
16. B   17. D   18. A

Chapter 15: Ideal Gas MCQs

1. C	2. B	3. D	4. C	5. A
6. A	7. B	8. A	9. B	10. B
11. C	12. A	13. D	14. B	15. C
16. C	17. C	18. C	19. A	

Chapter 16: Kinematics Motion MCQs

1. C	2. A	3. C	4. C	5. D
6. A	7. A	8. B	9. B	10. B
11. A	12. D			

Chapter 17: Kirchhoff's Laws MCQs

1. B	2. A	3. B	4. C	5. D
6. D	7. C	8. A	9. A	10. B
11. A	12. C			

Chapter 18: Matter and Materials MCQs

1. A	2. B	3. A	4. D	5. D
6. C	7. C	8. D	9. B	10. B
11. B	12. D	13. A	14. D	15. B
16. B	17. A	18. C	19. A	20. A
21. D	22. C			

Chapter 19: Mechanics and Properties of Matter MCQs

1. C	2. C	3. A	4. B	5. A
6. B	7. D	8. C	9. B	10. D
11. C	12. A	13. D	14. C	15. B
16. B	17. C	18. C	19. B	20. B
21. D	22. C	23. B	24. B	25. D
26. C	27. A	28. C	29. A	30. A
31. C	32. A	33. B	34. A	35. B
36. D	37. C	38. D	39. C	

Chapter 20: Medical Imaging MCQs

1. A	2. D	3. B	4. A	5. C
6. C	7. C	8. B	9. D	10. A
11. C	12. D	13. A	14. D	15. B
16. C	17. B	18. B	19. D	20. C
21. D	22. C	23. B	24. B	25. A
26. A	27. B	28. C	29. A	30. A
31. C	32. B	33. B	34. A	

Chapter 21: Momentum MCQs

1. C	2. B	3. A	4. C	5. C
6. A	7. A	8. B	9. B	10. D
11. B	12. C	13. C	14. C	15. A
16. B	17. D	18. D	19. A	20. D
21. B	22. A			

Chapter 22: Motion Dynamics MCQs

1. B	2. B	3. B	4. C	5. A
6. C	7. A	8. D	9. B	10. A
11. D	12. D	13. D	14. D	15. C
16. A	17. C	18. B	19. C	20. A
21. B	22. C	23. A	24. B	25. C
26. A				

Chapter 23: Nuclear Physics MCQs

1. C	2. C	3. A	4. A	5. A
6. B	7. B	8. A	9. A	10. B
11. D	12. C	13. D	14. B	15. D
16. C	17. A	18. B	19. B	

Chapter 24: Oscillations MCQs

1. B	2. B	3. C	4. C	5. A

6. D	7. D	8. B	9. A	10. B
11. C	12. B	13. C	14. B	15. C
16. C	17. A	18. B	19. D	20. B
21. A	22. A	23. C	24. B	25. C
26. C	27. C	28. A		

Chapter 25: Physics Problems MCQs

1. D	2. C	3. B	4. A	5. A
6. C	7. B	8. D	9. D	10. D
11. D	12. A	13. B	14. C	15. A
16. B	17. A	18. B	19. C	20. B
21. A	22. C			

Chapter 26: Physics: Waves MCQs

1. A	2. A	3. C	4. A	5. D
6. C	7. D	8. A	9. D	10. B
11. A	12. C	13. B	14. D	15. C
16. C	17. D	18. B	19. B	20. C
21. D	22. A			

Chapter 27: Quantum Physics MCQs

1. C	2. A	3. D	4. B	5. B
6. D	7. D	8. B	9. C	10. A
11. B	12. C	13. A	14. C	15. C
16. A	17. C	18. A	19. A	20. D
21. D	22. D	23. A	24. B	25. B
26. B	27. D	28. A	29. A	30. C

Chapter 28: Radioactivity MCQs

1. B	2. A	3. C	4. B	5. C
6. A	7. C	8. A	9. B	10. A
11. B	12. C	13. A	14. D	15. C
16. B	17. D	18. A	19. D	20. B

21. A	22. D	23. A	24. B	25. B
26. D	27. B	28. B	29. C	30. D
31. A	32. C	33. D	34. C	

Chapter 29: Resistance and Resistivity MCQs

1. B	2. B	3. B	4. B	5. B
6. C	7. D	8. B	9. C	10. C
11. C	12. D	13. A	14. A	15. A
16. C	17. A			

Chapter 30: Superposition of Waves MCQs

1. B	2. A	3. D	4. A	5. C
6. B	7. C	8. B	9. A	10. D
11. A	12. B	13. D	14. C	15. B
16. D	17. C	18. D	19. A	20. A
21. C				

Chapter 31: Thermal Physics MCQs

1. D	2. A	3. D	4. A	5. A
6. B	7. D	8. C	9. A	10. C
11. A	12. D	13. C	14. C	15. B

Chapter 32: Work, Energy and Power MCQs

1. D	2. B	3. C	4. C	5. A
6. B	7. A	8. C	9. B	10. D
11. A	12. C	13. A	14. A	15. B

Made in the USA
Monee, IL
23 August 2021

76309110R00114